Critical acclaim for Cynthia Moss and her *Elephant Memories*

"She conveys, more realistically than any other popular field account I have read, the day-to-day minutiae of collecting information. . . . Ms. Moss tells the story in a style so conversational, so free of academic jargon, that I felt like a privileged visitor riding beside her rickety Land-Rover as she showed me around the park."

Sarah Blaffer Hrdy
The New York Times Book Review

"Her accounts are laced with beautiful vignettes describing individual elephants as they are born, mature, and die. Moss offers a glimpse into a marvelously complex society where kinship is the thread connecting the lives of elephants."

Natural History

"Her findings are extremely interesting: about elephant love and breeding; about elephant family attachments, which seem to survive death . . . and startling discoveries about elephant communication."

The New Yorker

"Immensely readable and moving."

New York Daily News

ALSO BY CYNTHIA MOSS

Portraits in the Wild

ELEPHANT MEMORIES

Thirteen Years in the Life of an Elephant Family

Cynthia Moss

Fawcett Columbine • **New York**

In memory of the matriarchs in my family
unit, my grandmother Berthe Therese
and my mother, Lillian; and to the
present matriarch, my sister Carolyn.

A Fawcett Columbine Book
Published by Ballantine Books

Copyright © 1988 by Cynthia Moss

All photographs in the book were taken by Cynthia Moss
except where indicated.

Drawings by David Pratt
Diagrams by Liz Jarvis

All rights reserved under International and Pan-American
Copyright Conventions. No part of this book may be re-
produced or utilized in any form or by any means, elec-
tronic or mechanical, including photocopying, recording or
by any information storage and retrieval system, without
permission in writing from the Publisher. Published in the
United States by Ballantine Books, a division of Random
House, Inc., New York, and distributed in Canada by
Random House of Canada Limited, Toronto.

Library of Congress Catalog Card Number: 88-92233

ISBN: 0-449-90362-1

This edition published by arrangement with William
Morrow and Company, Inc.

Manufactured in the United States of America

First Ballantine Books Edition: May 1989
10 9 8 7 6 5 4 3 2

Acknowledgments

In a project covering more than 14 years, there have been many people and organizations that have contributed generously in a variety of ways. Without some of them the Amboseli Elephant Research Project would never have started and would not be continuing today. Others have contributed more indirectly but just as significantly.

I would like to start by thanking the Kenya government for allowing me and my colleagues access to their elephants, which they have wisely held in trust for all the world to enjoy and for a few of us fortunate ones to study. In particular I would like to thank the Office of the President and the National Council of Science and Technology for research clearance and the former Kenya National Parks and the present Wildlife Conservation and Management Department for permission to work and reside in Amboseli National Park. I have had a long and happy relationship with the Wildlife Department, and among the people that have been particularly helpful over the years I would like to thank Perez Olindo, who, as director of Kenya National Parks, first gave permission for the study to begin, and Daniel Sindiyo, director of WCMD, and David Mbuvi and Fred Pertet, deputy directors, WCMD, for their cooperation and interest in the subsequent years. In Amboseli itself I appreciated the kindness, hospitality, and generosity of the wardens, but in particular that of Joe Kioko and his wife, Christina, and Bob Oguya and his wife, Julia. The Kenya Rangeland Ecological Monitoring Unit has been the willing and very helpful local sponsor for the project and for this I owe my gratitude to David Andere, the director.

The project would never have begun without financial sponsorship and I am forever indebted to the support I have received from the African Wildlife Foundation. That support has been more than just financial; it has been a statement of faith in me throughout the years. Without AWF I would not have been able to carry on for as long as I have. Both the Washington office and the Nairobi field office have backed me up logistically and administratively, providing a firm base for the project since its onset. Those in the Washington office I would like to thank are the presidents over the years, John Rhea, Robinson McIlvaine, Robert Smith, and Paul Schindler, and vice-president Diana McMeekin, and secretary/treasurer Elizabeth McCorkle. In Nairobi my thanks go out to everyone in the office who has helped in so many ways—field directors Frank Minot, Robinson McIlvaine again, Robert K. Poole, and Sandy Price; program officer Rob Olivier; administrative assistants Ronnie Hunter and Sue Heather-Hayes;

accountants Netty D'Souza and Doreen McCullough; the staunch core of the office, Mary Metebo; and last but in no way less appreciated, Mukunjuro Gichoba and Timothy Gitau. They all made the sometimes stressful and tedious bureaucracy of running a project far smoother and easier by their assistance.

Other organizations and many individuals have also supported the Amboseli project financially. The New York Zoological Society generously funded my study for three and a half years and I would like to thank William Conway, George Schaller, and Archie Carr III. Individuals have in many ways been my most important sponsors; some have been loyal donors for years. They truly keep the study going and to them I owe a huge debt of gratitude. I will list some of the major donors with their foundations, if applicable, in alphabetical order: George A. Binney; Robert B. Glynn; Paul Hirsch; Charles Jackson, Jr.; Scott McVay (W. Alton Jones Foundation); Park East Tours; Richard Phippen (Hurdle Hill Foundation); Audrey Sheldon Poon (Merlin Foundation); Edmund Pratt (Pfizer Corporation); Denise and Sioma Schiff; Paul Simon; Mrs. Herman Wouk (Abe Wouk Foundation); and Richard Weatherhead (Midgard Foundation). Many others helped both financially and with publicity in response to my appeal to replace my Land-Rover and tent and, as I hope they all know, their efforts were greatly appreciated.

Support comes in many forms, much of it not financial. Many people in the United States, England, and Africa have helped me in immeasurable ways by their friendship and hospitality. In the United States I would like to thank the following people for their kindness over the years: Kate Coleman, Bob and Joan Donner, Bob and Pat Glynn, Mariana Gosnell, Harriet Huber, Michael Johnston, Jan and Carolyn Long, Barbara Maltby, Penelope Naylor, David Teitelbaum and Wendy Weil. In England I would like to thank Robert Hinde and Patrick Bateson for their hospitality at Cambridge University, Duncan Mackinder for sorting out my vast mounds of data with his outstanding computer skills and Richard Laws for his support. In Africa, I must first thank Iain Douglas-Hamilton for instilling in me a fascination with and concern for elephants. I would also like to thank the other elephant researchers in Africa and India who have shared their ideas and discussed their results with me. In Nairobi my friends have been exceptionally helpful and encouraging during my years of field research, data analysis, and book writing; I would like to thank Serafina Auma, Jared Crawford, Nani Croze, Virginia Finch, Andrew Hill, Robert Malpas, Sandy Price, Mark Riley, Kes Hillman Smith, and

Truman Young. Special thanks are due to Conrad Hirsh and Debbie Snelson for their companionship, care, and patience, and for all the delicious meals they cooked for me during the final year of writing.

The production of the book has been greatly aided by Jilo Quraishy of Photomural, who has developed and skillfully printed all my photographs since the study began. David Pratt and Liz Jarvis provided the lovely line drawings, endpapers, maps, and diagrams. I would also like to thank those who gave permission for their photographs to be used in the book: Harvey Croze, Joyce Poole, Marion Kaplan, Nicolai Canetti, and Athi-mara Magadi.

In Amboseli, where companionship, tolerance, and a sense of humor are all important, my friends have meant a great deal to me over the years. Their support and goodwill have been a very important element in keeping the long-term elephant project going. The lodge managers and staff have been helpful in many ways and I would like to thank David and Gillie Jackson, Gerry Sauer, Tamsin Corcoran, Patrick Njoroge, David Munguti, and Timothy Thuo of Amboseli Lodge; Tony Pascoe, Othoo Onyango, Jane Shabaan, and Morris Anami of Serena Lodge; and Mercy Sakawa of the Kajiado County Council. Others in Amboseli who have been good friends are David and Joyce Maitumo, Karenkol ole Musa, Kasaine Nkurupe, Daniel Nkurupe, Simon Ndungu, and Jama Sulieman. I also want to thank the women in the offices for their friendly assistance: Elizabeth Kitunka, Lucy Mtutua, Margaret Pertet, and Dorcas Saita. Fellow researchers living in Ol Tukai have provided stimulating discussion and friendship: Susan Alberts, Jeanne and Stuart Altmann, Jennifer Dillon, Glenn Hausfater, Lynne Isbell, Jeanne Meade, Amy Samuels, Jeff and Bev Walters, and Richard Wrangham.

Closer to home in the campsite I have shared many golden days and starlit evenings with people who have the same love of Amboseli. My good neighbors, who always came to the rescue in times of crisis or simply entertained me, have been Kay Behrensmeyer, Bob Campbell, Stuart Cunningham, Rob Foley, Chris Gakahu, Seido Hino and his Nippon TV crew, David Klein, Moses Kumpumula, and Vicki Morse.

In my camp I must first thank Masaku Sila and Peter Kumu Ngande for keeping my body and soul together. Without their hard work and skills the camp would not be the peaceful, beautiful home that it has become for me, and my appreciation for what they have done is boundless. I would also like to thank several assistants, students, and short-term visitors who have worked on the project: Claire Heffernan, Katy Payne, Titus Masaku, M. E. Mwange, Joan Riback, and Nellie

Sakawa. Norah Njiraini and Soila Saiyielel, who have joined the project recently have added greatly to the productivity of the elephant research and to the general well-being of the project and I am very grateful for their hard work and enthusiasm. My campmates, who have worked on nonelephant projects, have been very important to me, and through sharing the camp I have made some lifelong friends: Marc Hauser, Wes Henry, Cynthia Jensen, and Robert and Dorothy Seyfarth have all been helpful, amusing, and stimulating companions.

Last but most important, I want to thank my colleagues on the Amboseli Elephant Research Project. Jonah Western and Harvey Croze were instrumental in getting the project going and I appreciate their faith in me and support from the beginning. Colleagues Joyce Poole, Keith Lindsay, Phyllis Lee, and Sandy Andelman have all been close friends and companions. They have given me a great deal of assistance, for which I can never thank them enough. They have helped in field work, in collecting long-term records, in maintaining the camp, in analyzing data, in writing papers, in reading manuscripts, and in many other ways. Sharing the camp brought us close together, but sharing the elephants added another dimension to our relationship. No one who gets to know the Amboseli elephants as individuals is untouched by that knowledge and it has bound us irrevocably. For me it is a very good feeling to know that there are others who care about the elephants as I do, who will work to ensure that there is a viable future for them, and who, if there is a fight to save Amboseli's elephants, will be beside me. For that I thank my colleagues with all my heart.

Contents

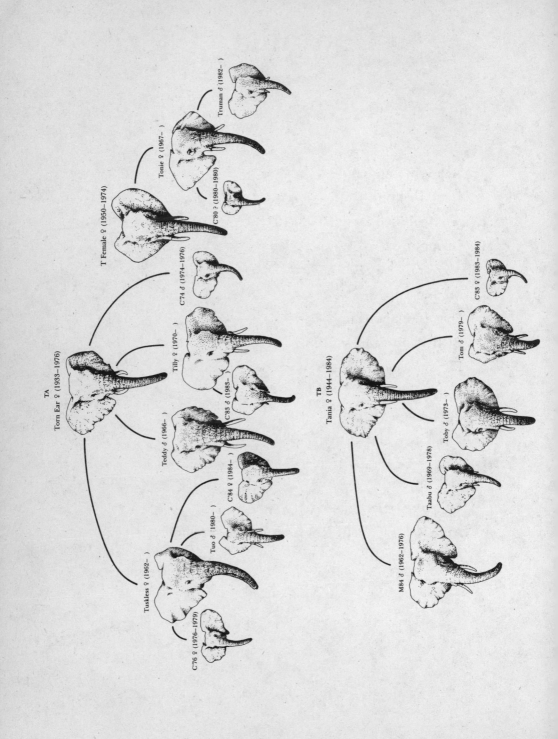

TA
Torn Ear ♀ (1933–1976)

T Female ♀ (1950–1974)

Tonie ♀ (1967–)

Truman ♂ (1982–)

C'80 ♀ (1980–1980)

C'74 ♂ (1974–1976)

Tilly ♀ (1970–)

C'83 ♂ (1983–)

Teddy ♂ (1966–)

C'84 ♀ (1984–)

Tuo ♂ (1980–)

Tuskless ♀ (1962–)

C'76 ♀ (1976–1979)

TB
Tania ♀ (1944–1984)

Tom ♂ (1979–)

C'83 ♀ (1983–1984)

Toby ♂ (1973–)

Taabu ♂ (1969–1978)

M84 ♂ (1962–1976)

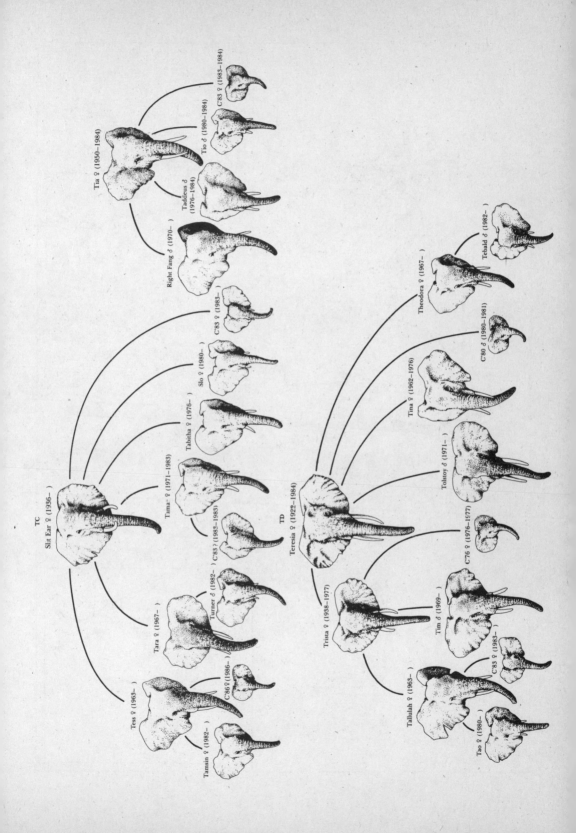

TC

Slit Ear ♀ (1936–)

Tia ♀ (1950–1984)

Right Fang ♂ (1970–)

Taddeus ♂ (1976–1984)

Tio ♂ (1980–1984)

C'83 ♀ (1983–1984)

Tama ♀ (1971–1983)

Tabitha ♀ (1976–)

C'83 ♀ (1983)

Slo ♀ (1980–)

Tara ♀ (1967–)

Turner ♂ (1982–)

C'83 ♀ (1983–1983)

Tess ♀ (1963–)

Tamsin ♀ (1982–)

C'86 ♀ (1986–)

TD

Teresia ♀ (1922–1984)

Theodora ♀ (1967–)

Tebald ♂ (1982–)

Tina ♀ (1962–1976)

C'80 ♂ (1980–1981)

Tolstoy ♂ (1971–)

Trista ♀ (1938–1977)

Tim ♂ (1969–)

C'76 ♀ (1976–1977)

Tallulah ♀ (1963–)

Tao ♀ (1980–)

C'83 ♀ (1983–)

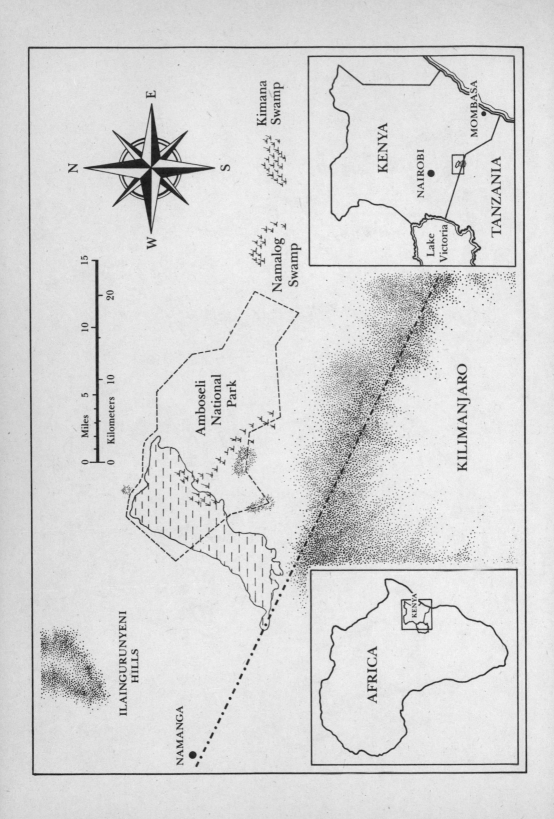

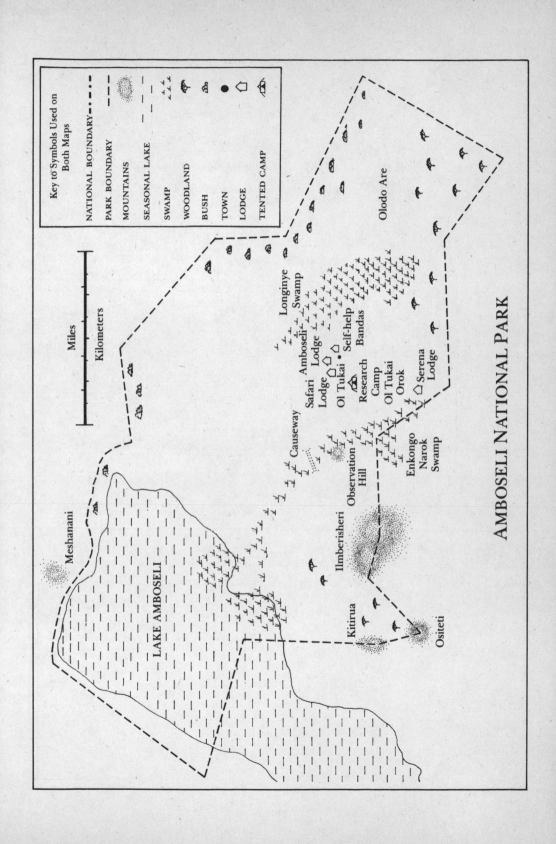

AMBOSELI NATIONAL PARK

Key to Symbols Used on Both Maps

NATIONAL BOUNDARY
PARK BOUNDARY
MOUNTAINS
SEASONAL LAKE
SWAMP
WOODLAND
BUSH
TOWN
LODGE
TENTED CAMP

Miles

Kilometers

Olodo Are

Longinye
Swamp

Safari Amboseli
Lodge Lodge
Ol Tukai
Research
Camp

Self-help
Bandas

Ol Tukai
Orok

Serena
Lodge

Causeway

Observation
Hill

Enkongo
Narok
Swamp

Ilmberisheri

Meshanani

LAKE AMBOSELI

Kitirua

Ositeti

If the elephant vanished
the loss to human laughter,
wonder and tenderness would
be a calamity.

— V. S. Pritchett
reviewing *The Roots of Heaven*

Introduction

There is a small national park called Amboseli at the base of Kilimanjaro in southern Kenya. In its 150 square miles live a variety of wild animals: lions, leopards, cheetahs, rhinos, giraffes, buffaloes, zebras, wildebeests, and elephants. The wildlife has lived there for several thousand years; I have lived there for most of the last 14 years, carrying out a study of the elephants.

Elephants are very special animals: intelligent, complicated, intense, tender, powerful, and funny. I consider myself immensely fortunate to have spent so much time with them. I have followed the lives of Amboseli's elephants through droughts to periods of superabundance, through times of heavy poaching and great losses to times of peace and relative security. I have watched them give birth and I have watched them die. I have observed young females reach sexual maturity and mate for the first time and young males leave the security of their families and strike out on their own. I have seen the grand old matriarchs leading and defending their families and

I have also seen them lose all dignity and run around in play with their tails curled up over their backs and a wild glint in their eyes.

After so many years with elephants I wanted other people to know them, enjoy them, and care about them as I have. I wanted to share my knowledge and experiences, and therefore I have written this book. I think it is important to point out that this is not a tale of adventure about human hardship and bravery in the face of Africa's heat and dust and large, dangerous animals; in other words it is not about how I survived alone in the bush with 600 wild friends. It is about the elephants' lives, how *they* survived or succumbed to droughts, poachers, Maasai warriors, disease, injuries, tourists, and even researchers. It is about their families, their "friends," their mates, and their offspring; and their good times and bad times through the seasons and the years. I have merely gone along with them — a spectator at the banquet or a witness to less happy events — in the role of voyeur, not as a participant. I am there, of course, but only as observer, interpreter, scientist, conservationist, philosopher, and sometimes just as a lover of elephants who puts aside the tools of the trade and simply enjoys being with them.

The book concentrates on the lives over the last 13 years of one extended family of elephants. In following the histories of these individuals from the time I met them to the present I hope to share with my readers the richness and complexity of elephant life.

The chapters run chronologically, starting in 1973, and each consists of three parts. At the beginning of each is a semifiction-alized scene in the lives of the family I have chosen. The scenes are based on a combination of actual events that I have witnessed, circumstantial evidence such as the disappearance or injury of an animal, and purely made-up material based on my knowledge of elephants. For example, in Chapter IV I describe the mating behavior of two of the females in the family. I saw one of the females in close consort with the male I mention for a few hours on one of the days she was in estrus and saw her chased by other males on another day. I did not see the second female in estrus but I know that she had to have mated around that time because she gave birth 22 months later. I have seen dozens of matings and have watched and recorded the behavior of well over 100 females in estrus. With these data I have been able to work out the typical course of events for both young and more experienced females in estrus. In another example, in Chapter IX I did not actually see the death I describe, but I have watched sick and dying elephants, and I have seen family members trying to lift and hold up a sick companion. Thus what I have done is taken basic events in the life of the family that I know had to have happened and filled in the blanks based on thousands of hours of observation of these and other elephants. I feel justified in creating these scenes because most people cannot fill in those blanks, but by my doing so I hope to bring my readers close in with the elephants, watching and imagining with me those events in their lives.

In most of the chapters the introductory scene is relatively short. The remainder of each chapter consists of two elements. One is the theme, such as Mating, Social Organization, or Population Dynamics. I discuss these topics in detail, often using examples from the family I have concentrated on. The third part relates observed events and developments in the lives of the family and usually ends with a list of family members at the end of the year that the chapter has covered. Sometimes the themes and events are interwoven, and thus there is not a rigid structure to the second and third parts of each chapter.

Like a family saga or a soap opera, there are many names to remember and the relationships are sometimes complex. As an aid to keeping track of who is who I have included a family tree. There are also two maps to keep one from getting lost among the Maasai names and a diagram of the multi-tiered social organization.

I have always said that watching elephants is like reading an engrossing, convoluted novel that I cannot put down but I also do not want to end. I hope in this book I have conveyed something of that feeling and that my readers will understand why elephants are so fascinating and so special and why they are deserving of respect and consideration.

CYNTHIA MOSS
Nairobi
March 1987

Chapter I

An Amboseli Day

1973–1975

In the early morning light, the large gray forms moved slowly, deliberately, and nearly silently along a well-worn path amongst lava boulders and small thorny *Balanites* trees. They walked in single file with a large female with long, upcurved tusks at the lead, many calves of varying sizes and ages along with more adult females in the center, and a straight-tusked, ancient, but immensely dignified female at the rear. Behind them rose the overwhelmingly dominant feature of the landscape — the 19,340-foot snow-capped peak of Kilimanjaro. The elephants headed north, away from the mountain; they walked without stopping to rest or feed, as if they had an appointment at a prearranged place.

Eventually they emerged into a habitat of open pan with no trees and little other vegetation. Here the quiet *sluff, sluff* of their feet kicked up the fine alkaline dust and their outlines became hazy. They did not linger here either, but continued on until abruptly the dry, bare ground gave way to lush, green vegetation interspersed with open water dotted with and sur-

21

rounded by myriad water birds: ducks, geese, ibises, herons, cranes, plovers, jacanas, and many more. Other animals gathered along the edge of the swamp: zebras, wildebeests, reedbucks, waterbucks, and buffaloes. The elephants kept on their course, ignoring these smaller species, who stepped discreetly out of the way.

On reaching the swamp, the elephants spread out. Some began to feed immediately while still walking, snatching up large mouthfuls of the dark, green *Cynodon dactylon* grass, others continued on a straight course, and all arrived together at a channel of clear running water. Each of the adult females dipped the tip of her trunk into the stream, sucked the water up into her trunk, lifted her head up and back, placed the trunk in her mouth, and let the water flow back down her trunk and down her throat while swallowing at the same time. Most of the calves used the same technique, but one young calf was not yet adept with his trunk and he knelt down and drank with his mouth.

When they finished drinking some went on into the swamp and started feeding in earnest, while others walked over to a depression where a mud wallow had formed. The adults splashed themselves with the dark, glutinous mud by first picking it up in the curve of their trunks and then flinging it onto their chests, backs, sides, and heads. The mud hitting their bodies made a sharp, wet slapping sound. Soon they turned from a light gray to mostly glistening black. The calves, showing less restraint, waded right into the wallow and flopped down on their sides and then writhed and wriggled in the mud until one side, including head, ears, and eyes, was completely covered in mud. Then they sat up and flopped over to the other side. Young calves took the opportunity to climb on the older ones as soon as they were down, and in the process, got covered in mud themselves. Eventually there was a great heap of youngsters completely filling the mud wallow. Slowly they disentangled themselves and struggled out of the mud wallow, slipping and sliding up the small bank.

The calves joined the adults, who were now moving deeper and deeper into the swamp. There the elephants found grasses, sedges, including papyrus, and succulent creeping herbs. They ate steadily, building to a rhythm: first twisting the trunk around a bunch of vegetation, pulling to one side and ripping the bunch free, then placing it in the mouth, and immediately reaching for more as they chewed. Some kinds of plants came out by the roots and with these the elephants bit off the part that was palatable and let the other drop. The small calves found less that they could handle, only young shoots here and there and tender creepers. These younger calves also had some trouble trying to follow their mothers through the dense vegetation and deep water and at times they had to swim from one clump of vegetation to the next.

The elephants fed continuously until midday, when they moved to higher, dry ground, where they found a bare patch of dusty soil. They scooped the dust up in their trunk tips and blew it out over their heads and backs. Now they gathered together in a tight group, nearly a circle, standing close together, some touching. One female rubbed her head against the shoulder of the old, straight-tusked female. A calf leaned and rubbed against the leg of his mother. First one calf lay down, then three more subsided to the ground and lay flat on their sides. The females' heads hung down and their trunks became limp and stretched out until the tips touched the ground. Two of the females rested their trunks on their tusks. One simply draped her trunk over her left tusk. The upcurved female neatly curled her trunk like a snake and rested it on both her tusks. A half-grown calf placed his trunk on the back of a sleeping calf. All became quiet and breathing deepened.

The elephants slept for about 40 minutes, the calves soundly, but the adults much more lightly, occasionally opening their eyes, gently flapping their ears, or swishing a trunk or tail at an annoying fly. A young adult female was the first to show signs of stirring. She stepped forward a few feet and started dusting. Then the calves stood up and synchronously most of the members of the group defecated and urinated. Still they stood there as if waiting. Then the ancient female made a very long, soft rumbling sound, raised and flapped her ears against her neck and shoulders, letting her ears slide down with a rasping sound, and set off. This was the signal they had been waiting for and all the others followed after her.

They traveled north, forming a column two or three abreast. Once again they seemed to have a purpose in their movement and direction. Some of them raised their trunks in the air, sniffing the wind. They headed toward a channel of the swamp where the water continued to flow above ground. Feeding near the channel were two more groups of elephants. The original group moved steadily in their direction. When they arrived at the first of the two stationary groups there was a discreet reaching of trunks in each other's direction by some of the members, but the upcurved female and the old female ignored these elephants and walked on past toward the second group. When they were about 50 yards from this group, the up-curved female rumbled — a different, higher-pitched, and louder rumble than the signal the old female gave after they woke. This sound produced an instant reaction in the new group. The members raised their heads, lifted and spread their ears, and produced loud, throaty rumbles. At the same time the elephants in both groups began to secrete a clear liquid from the temporal glands on the sides of their faces. More answering rumbles came from the original group and then both started striding rapidly toward each other. When they were 20 yards apart they broke into a run and came together in a turmoil of earflapping, rumbling, screaming, trumpeting, clicking of tusks together, entwining of trunks, spinning and backing, and urinating and defecating. Their temporal gland secretions were streaming down their faces and they reached their trunks toward each other's glands. The upcurved female pushed through the milling calves and went straight to a large female with a big tear out of her right ear. They lifted their heads together and clicked tusks while entwining their trunks and rumbling deeply.

The elephants continued to greet one another, with accompanying loud vocalizations, for another four to five minutes. When they calmed down the adults and older juveniles started to feed. The younger calves, still stimulated by the interactions, investigated each other for a while longer. Eventually, they too began to feed again. The two groups had now merged into one and all fed together peacefully. They moved into the deeper part of the channel and fed in this area for another two hours.

Around 4:00 a huge lone bull elephant, carrying his head high and his chin tucked in, appeared from the east. His temporal glands, also located midway between his eye and ear on each side of his head, were streaming with a thick, viscous fluid. The sheath of his penis was a greenish color and continuously dripped urine. He gave off an extremely strong, sharp odor. When he was still well over a hundred yards away the females in the group raised their trunks and sniffed in his direction. They stopped feeding, rumbled, and warily watched his approach. He continued to come, and as he got closer he slowed his pace, lowered his head slightly, and nonchalantly draped his trunk over one tusk. The females visibly relaxed, and when he arrived in the group, they rumbled excitedly and some of the younger females turned and backed in toward him. The bull went to each female in turn, and placed the tip of his trunk on her vulva, which hung down low between her hind legs, facing the ground. If she had urinated he touched the urine with the tip of his trunk. With most of the females he quickly moved on; with a few he smelled the vulva or urine for longer and then placed his trunk in his own mouth and stood very still as he analyzed the odor and taste. After he had tested all of the females, he started to feed, but after about half an hour he left and moved on to the other group in the area, where he also inspected the females. Here too he found nothing to interest him and he moved on, heading out of the swamp to the west.

Toward the late afternoon the new aggregation of females and calves began to orient toward the south and slowly amble in that direction while they fed. They followed the course of the channel for a while but eventually came up onto higher ground and out into the grasslands. They now joined the second group that had been feeding nearby, but they exchanged no greetings. They became a single broad front of elephants moving and feeding through the grass. As the sun was setting they reached a small grove of young *Acacia xanthophloea* trees. Some of the animals fed on these, carefully removing the thorny branches, and then manipulating them with trunk, tusks, and feet to remove the bark from the larger branches or to bend the thorns in one direction before placing the smaller branches in their mouths.

While the adults were more or less stationary, the calves took the opportunity to play. The young juvenile males from all three groups came together to spar vigorously in pairs. Two faced each other, lifted their heads, and came gently together nose to nose. In this position they felt each other's heads and mouths, with their trunks winding around and through their tusks. They began to push against each other, moving backward and forward until one broke away and backed off. They came together again and pushed until one was shoved onto his backside. They separated, backed up, and this time ran at each other and met with the rich chunking sound of ivory against ivory. Finally one broke away and turned and ran with the other following. The first one whirled around, lifted his head, and spread his ears and his pursuer stopped short, watched the other for a moment, then picked up a stick in his trunk and threw it in the air. The other young male lowered his head and approached and the sparring began again. Several pairs of young males sparred in this way, while the younger calves imitated them, chased one another, climbed on each other, or simply ran around for no apparent reason. They squealed, bellowed, and trumpeted, but the adults paid them no attention whatsoever.

Slowly, one by one, the adults stopped feeding and moved away from the trees. They stood at the edge of the small grove orienting south. Once again they waited until the old female rumbled before setting off. Then they formed a line and headed toward the mountain. The younger calves joined their mothers, but several older calves continued to play. Suddenly finding themselves abandoned, they broke off from sparring to catch up with the others. They ran with a loose, floppy gait, shaking their heads from side to side, letting their ears flap against their necks, and curling their tails up high over their backs, all the while trumpeting with a loud, pulsating, nasal sound. As they neared the group one of the young males whirled and playfully attacked another and they began a new sparring match. Once again they got left behind and once again they ran trumpeting to catch up. The young animals ran and played, and ran and played all across the pan. When they reached thicker vegetation they started to bash through bushes, beating them with their heads and tusks. Even some of the young adults joined in this game and the sounds of their trumpets and screams rent the otherwise quiet evening.

Suddenly, all together, as if by signal, the elephants were quiet and still. They cautiously lifted their trunks, smelled the

air, and then took off at a very fast run, in tight formation, with
the calves close to their mothers and the larger adults at the
front and rear. They ran fast and silently with no trumpets or
screams. Nearly a half a mile away, silhouetted against the last
light in the western sky, three Maasai warriors moved across the
plain with their characteristic ground-eating, loping gait. Over
their shoulders they carried six-foot-long spears.

The elephants ran for three quarters of a mile before slowing
to a walk. They stopped, still bunched together, turned, and
faced in the direction from which they had come, smelling the
wind. One or two gave a sharp toss of the head, which made
their ears crack like canvas sails. Several of them extended
trunks toward each other, and most of the mothers reached for
their babies and felt them with their trunks. After a while they
relaxed.

They were back in the area of lava and *Balanites* trees that the
original group had passed through in the morning. Soon the
aggregation broke down into subgroups and spread out to
feed, looking for grass and small shrubs and herbs growing
among the rocks. After a while the upcurved female and the
ancient female and their calves and associates were on their
own once again. Even they were split up into subunits as they
continued to feed and move slowly toward the mountain. Every
ten minutes or so one of the adult females rumbled a low
contact call and listened for an answer from the other group
members keeping in touch. By midnight they had covered only
a few miles. One by one they stopped feeding and reformed
their group, gathering around the two big females. First the
young calves, then the older ones, and finally the adult females
all lay down and went to sleep. In the moonlight they looked
like huge gray boulders, but their deep, peaceful snoring belied
the image.

Around 4:00 A.M. they began to stir and struggle up onto
their feet. They urinated, defecated, rubbed their eyes, and
scratched themselves with their trunk tips or picked up a stick
and used it to scratch their bellies or between their front legs
where the huge elephant ticks lodged. One female found a
convenient tree stump and rubbed herself against it in an
ungainly posture. There seemed to be no hurry to go anywhere,
but eventually the upcurved female walked off and started to

feed and others followed. Their general movement was to the northwest, away from the mountain again.

The elephants fed until dawn and then started striding out more deliberately until, like the morning before, they formed a single file and walked steadily along a well-worn path heading toward the swamp.

I first met the big upcurved female and the ancient female on September 9, 1973. I found them in one of the woodland areas of Amboseli called Ol Tukai Orok, which means "place of the dark palms" in Maa, the language of the local Maasai people. I was in Amboseli for a few days trying to photograph as many elephants as possible to build up a recognition file of the population. I had not seen these animals before and was pleased to find them in a fairly open area where they were feeding and moving slowly. That meant I could take pictures. I was also pleased that they appeared to be very tolerant of my vehicle.

First I counted the group several times and got 11, then 12, then 13 three times in a row. It is more difficult to count elephants than one would imagine. They are large animals and so one assumes that one would be unlikely to miss any, but it is because they are so large that it is easy to miss the ones that are hidden behind the bulk of the nearer ones. After three counts of 13 in different combinations I was satisfied that there were "13 exact" as I wrote in my field notes.

I then tried to work out the age and sex structure: There appeared to be four medium-to-large adult females (anywhere from 20 to over 60 years old), two young adult females (about 11 to 19 years old) and seven calves (newborn to about ten years old) of various ages and both sexes. I wrote this down and started photographing. Ideally I wanted to photograph each animal's head and ears from the left side, right side, and head on. This is easier said than done, and typically I had to be content with the left sides of some, the right sides of others, and head-on shots of still others. The important thing was to keep very accurate notes of each photograph so that I did not assume that the right ear in picture #33 went with the left ear in picture #35 when they were actually two different animals. My notes looked like this:

Right side upcurved female — 33
Right side 2½ yr calf with her — 34
Left side ≈ 15 yr old F with a big V notch out of each — 35
    ear
Left sides large female with long straight tusks — 36
    and her 2 yr calf

I had learned these techniques from Iain Douglas-Hamilton, a pioneer in the use of individual recognition in wildlife research. He studied the elephant population in Lake Manyara National Park in Tanzania and was the first person to use photographs of elephants so he could recognize large numbers of individuals over and over again. By the end of his four-and-a-half-year study he knew almost every individual in a population of over 500. I worked for Iain for eight months in 1968 and for shorter periods in 1969, 1970, and 1971, assisting him in field work and analysis of data. It was at Manyara that I got my "eye" for elephant ears, some imprinted so indelibly on my mind that I can still remember them to this day.

It is primarily the ears that I use to identify individuals. First of all, African elephants' ears are big and conspicuous. Second, they are rarely smooth along the outer edges. Elephants usually have holes, nicks, and tears on the edges of the ears. In addition, the veins in the ears are often prominent and the pattern they form is unique, as accurate a means of identification as human fingerprints and a lot easier to see. Thus if an elephant has completely smooth edges to its ears with no holes, then the vein patterns can be used. In any case a combination of nicks and holes and vein patterns guarantees accurate identification once a good photograph has been taken.

I also use a variety of secondary characteristics. Tusk length, circumference, shape, and configuration all help to make an elephant look individual. However, tusks often break and when they do the general appearance of the elephant can change markedly. Other oddities such as scar tissue, warts, growths, injured legs and trunks can also be used as secondary aids, but these too can disappear or change. In the end, I always rely on the ear pattern, which changes only gradually over the years and then usually only one section at a time. The veins, of course, remain the same throughout life.

I discovered in Manyara that once I got to know an elephant I was not using any one of these characteristics in my initial, immediate recognition of the elephant. Instead I got to know the whole elephant: its size, shape, posture, the way it moved, stood, or carried its head. I would see an elephant from over 100 yards away and say to myself, that's so-and-so, and then when I drove up to it I would double-check its ears. It is a bit like recognizing a human friend who is walking away from you on the other side of the street. You know that friend by a combination of many features, some of which you would not even be able to describe. When you call out and the friend turns around, you are positive who it is. When I get closer to an elephant and I see its ears, I too can make a positive identification.

When I started my own study of elephants in Amboseli National Park in Kenya in 1972, I was fortunate to have had the experience of working with Iain, who not only taught me how to identify elephants but also how to work around them, how to estimate their ages, and how to study their social organization and behavior. I was doubly lucky in that, in the first two years of the Amboseli study, I had, as a partner, Harvey Croze, who had recently completed an ecological study of the elephants of the Serengeti National Park in Tanzania. He had just become a lecturer in the zoology department of the University of Nairobi and hoped to continue his elephant work in Kenya. I wanted nothing more than to study elephants again after my experiences in Manyara, and when David Western, an ecologist who had worked in Amboseli for several years, suggested that someone should study the elephants in Amboseli, Harvey, whom I knew slightly, and I decided to join forces and work together.

In the first two years of the study Harvey and I went to Amboseli separately or together whenever we could get away from our various jobs and responsibilities. Harvey managed to obtain some funds for the project from the university and from the East African Wild Life Society. I used my own money that I had saved. It was by necessity a part-time study. Our primary goal during those years was to complete a recognition file on the elephants. Once we knew individuals we could gather data on social organization, daily activities, ranging patterns, and reproductive behavior.

On that first occasion when I met the upcurved female and the old female, I was in Amboseli on my own for a few days. Two weeks later I was back with Harvey. On the morning of September 23 we found the upcurved female and some other elephants that we had not seen before. They were just about to move into the deep part of one of the swamps and we barely had time to count them and take a few photos. This swamp, called Enkongo Narok, which means "black eye" in Maa, is actually a relatively narrow water course with thick vegetation growing in water up to eight feet deep. This does not deter the elephants, however. They plunge right in up to their ears and feed in the deep water for hours. On this morning, the elephants quickly disappeared among the reeds and we had to leave them. By that date in 1973 we knew it was a typical pattern for the elephants to come into the swamps in the morning, feed there for most of the day, and then move out into the bushland in the late afternoon or evening. We went back at three o'clock in the afternoon and were lucky enough to find the group just coming back out onto the shore.

Now we had ideal conditions for taking photographs of the individuals. A large female with a V-shaped notch out of her right ear with a small flap hanging down from the lower part of the notch was feeding nearby on a *Suaeda monoica* bush. Near her was a young tuskless female, about eleven years old. We had already seen a few other tuskless females in the Amboseli population and were curious to know how many there were and if tusklessness or one-tuskedness ran in families. Next we photographed a very pretty female with almost completely smooth ears. I use the word *pretty* here with no hesitation, because anyone who watches elephants on more than one or two occasions can easily see that some elephants are better-looking than others, at least to the human eye. It has something to do with proportions: the size and set of the ears; the length and width of the trunk in relation to the head; the size, shape, and carriage of the head; the length of the legs; the curve of the back; the general body shape; and perhaps most striking, the particular configuration of the tusks. Apparently it is not a totally subjective impression. Inevitably two or three people watching a herd of elephants together will agree that that one over there is "beautiful," this one is "pretty," that one is "funny-looking," or even this one is "ugly." So in the beginning, when trying to sort out a group of elephants, one cannot help using these overall impressions as an aid in remembering who is who. With the pretty female was a funny little calf who had one left tusk and a tiny, stunted right tusk that looked more like a tooth than a tusk. On the spot he was called "Right Fang" by Harvey.

Next we photographed a very handsome, medium-sized female with relatively smooth ears. She had with her a young male calf only a few months old. As soon as this female got onto dry ground her calf hurried forward to his mother's teats. She put one leg forward and he suckled vigorously for a minute and a half on her right side, then quickly ducked under her chest and suckled on the left side for another 15 seconds until she moved. They had been in the swamp for over four hours and he had probably not been able to suckle during that time.

The upcurved female I had photographed two weeks before now emerged from the swamp, followed by several calves. We took more photographs of her and noted that she had a very neat slit in the middle of her right ear. I drew a picture of it in my notes:

I expected to see the old female with the long straight tusks and the young female with the two V notches, but when all the elephants had come out of the swamp we could see that they were not present.

After the elephants had fed for a short while on the bushes at the edge of the swamp, we heard three low rumbles and then they started moving west along the edge of the swamp. We counted 18 animals, including an adult male who surreptitiously tested the females as he walked along.

The elephants came to a causeway, which had been built

across the Enkongo Narok water course so that vehicles could pass to the western side of the park. Elephants could easily cross the swamp by moving through the water, but sometimes they used the causeway. On this day, it appeared that the elephants decided to avoid getting wet in the swamp again. According to my field notes, at 5:25 they walked up onto the road and started down the narrow track, which had low walls on both sides to prevent cars from driving into the water. Coming along the causeway from the east side was a lioness, followed by a tourist vehicle. The lioness was about halfway across the causeway when she noticed the elephants. She stopped short and started to turn back, but the vehicle was in her way so she turned again and started trotting toward the elephants, who got alarmed and moved back. Not for long, however, The torn-eared female came forward with her head high and her ears spread and the lioness instantly crouched down with ears flat against her head. The torn-ear was now joined by the tuskless young female, a calf, the slit-eared upcurved female, and the adult bull. They stood their ground facing the lioness. My notes read:

1732 Torn ear starts moving toward the lioness and car. Lioness gets up and retreats around car. Tuskless and younger calf follow, then bull, then slit ear. Others except female with first year calf come up on causeway.

1740 The elephants mill about — car still on causeway unable to cross.

1743 Car backs up and [only then] elephants turn around and come back off causeway on our side.

1745 Bull last off.

1746 1 min later the lioness returns down the causeway with two cars following her. She turns upstream. The elephants are downstream.

The elephants had held up the lioness and traffic for over 20 minutes. The torn-eared and slit-eared females were endeared to Harvey and me from that point on.

We managed to photograph several groups of elephants on that trip to Amboseli. When we returned to Nairobi we spent the following weeks developing films, making contact sheets, and then choosing the best pictures and making prints of the individuals. For some reason I loved this part of the work. It was like a huge puzzle, trying to recognize individuals, putting left ears with right ears, assigning calves to this female or that, determining the exact number and composition of each elephant group. Harvey and I would spend hours in the darkroom developing, printing, washing and drying, then I would take the photographs home and pore over them using a magnifying glass and with the aid of our notes try to sort out all the ears. It was immensely satisfying when a group started to take form and I was sure I knew who its members were. If we were very, very lucky and saw a group completely on its own two or three times with good conditions for taking photographs, we could feel confident that we knew all the adult females in the group and had a good idea of how the various youngsters fit in.

Since we both had studied elephants before, Harvey and I were familiar with their basic social organization. Wherever they had been studied, it had been found that elephants lived in groups, commonly called "family units," which consisted of related adult females and their immature offspring, ranging from newborns to calves up to about ten years old. Each family was led by the oldest female, the matriarch. Male calves left the family on reaching sexual maturity at anywhere from 10 to 15 years old. Female calves stayed in the family and began to breed at around 11, producing their first calf at about 13. Adult males associated with the females and calves only infrequently. In addition, Iain Douglas-Hamilton had found that family units often had special relationships with certain other family units in the population and they spent a great deal of their time together. He called the families making up such an association a "kin group."

Thus Harvey and I knew more or less what to expect in terms of basic social organization. However, a careful scientist tries not to approach a subject with preconceived ideas, so we rigorously took down data on which elephants were present each time we saw a group and did not just start lumping elephants into family units because other people had found that social system. But the accumulated data on who spent time with whom soon revealed that the Amboseli elephants did, indeed, also live in family units.

A family unit in Amboseli averaged nine to ten in size and the members formed a tight-knit coordinated group. When we found a family group it was almost invariably synchronized in its activities. That is, all the members would be feeding, or walking, or resting, or drinking or mud wallowing. It was rare to find members doing several different things. Also the individuals were usually within a few yards or less of each other. A female and her youngest calf were frequently within a few feet of each other and often touching, while her older calves were never far away. When the group was feeding, the individuals spread out a bit; but while traveling or resting or standing, they were closely bunched. They appeared to be very tactile animals, often touching each other with trunks, or leaning or rubbing on one another.

The social bonds among family-unit members were obviously very strong, and one of their manifestations was the frequent greeting of one another. Often after they had been spread out feeding and the group coalesced, individuals would greet one another with a special posture and rumble. The greeters would first raise their heads, lift their ears and spread them, tuck their chins in, and then rumble loudly and throatily while flapping their ears. Sometimes greetings were more excited and intense and they gave us an idea of the degree of attachment between individuals and eventually between family units.

Harvey and I recorded that certain families spent more time with each other than with other families in the population, just as Iain had described in Manyara. When these families were together they moved in a coordinated manner and synchronized their activities. When they rested, they all gathered together in a tight-knit bunch, touching and leaning on one another. When they came together, they performed the greeting ceremony. Eventually I decided to call the association between the families that seemed to have special relationships a "bond group." Although I suspected that the families making up these groupings were one extended family, I did not definitely know if the members were related, so I decided to use *bond group* rather than Iain's term *kin group*.

Once we started working out the families, we had to devise a designation system for the individuals and family units. Numbers would be simplest, but experience had taught us that numbers were hard to remember once one had several hundred. "Now is she F 121 or F 132?" We knew that we would eventually be putting data into a computer so we decided to give each of the females a three-letter code derived from an easily rememberable name. As we photographed more and more individuals and met new groups, we started to give each tentatively determined family a letter of the alphabet. So the first family we ever photographed back in September 1972 was called the "AA" family. All the females in that family were given names starting with the letter A: Annabel, Alyce, Amy, Alison, Agatha, Amelia, Abigail, and Wart Ear. Wart Ear? Well, yes, we hit a snag. We had been calling this female Wart Ear (because she had a large wart at the top of her left ear) ever since we first met her family, and by the time we felt confident enough to assign the family a letter and name the females, her name had stuck. We took the first three letters of the names and used them as their computer codes: ANN, ALY, AMY, ALI, AGA, AME, ABI, and WAR. Alyce was spelled with a *y* instead of an *i* so she would have a different code from Alison.

We tried to choose common English and European names for the elephants because they were the easiest for us to remember. I am very thankful that we did so because today, with over 500 elephants named, it would have been very difficult to remember any obscure names. To this day I cannot remember who is who among four young males with Russian names in the "V" family. Vladimir, Vostok, Vasily, and Vronsky are forever confused in my mind and I *always* have to look them up before writing them down. Also, and this is a key, I was not the one who named them — they were named by colleagues later in the study. Naming is a fascinating phenomenon and a surprisingly powerful process. Somehow by naming something one possesses it, almost creates it. At the same time one feels a closer relationship to that thing. I did most of the naming in the early days of the study, but Harvey named a few of the animals, and although he has had little to do with the study in the last ten years I still think of Tania, Filippa, Justine, and Jezebel as "Harvey's elephants." When Filippa died in 1982 it was Harvey I thought of and wanted to tell.

Purists might say that it is dangerous naming animals in a scientific study because one might associate names with people one has known and then impose that person's characteristics on the animal. I know that none of us who have worked on the project has found this. An elephant is so much its own being that it soon overshadows any association with a name. I, however, have the opposite problem now. When I am introduced to a person named Amy or Amelia or Alison, across my mind's eye flashes the head and ears of that elephant. I am also constantly looking for new names for elephants. I have been through four "names for babies" books and am now on a dictionary of saints. When I am introduced to someone with an unusual name that I have not used yet, I get excited and ask her to spell it for me. Books sometimes help. Harvey and I were both reading Márquez's *One Hundred Years of Solitude* in 1974 and found the names particularly suitable for elephants, and so we have Remedios, Ursula, Renata, Pilar, and Aurelia (the feminine form of the main character's name).

It took several more trips to Amboseli before we were tentatively able to sort out the families with the torn-eared female, slit-eared upcurved female, and old straight-tusked female. Repeated sightings of the same individuals led us to conclude that the torn-eared female, whom we named Torn Ear and estimated to be about 40 years old, was the matriarch of a small family which consisted of four animals, herself and three offspring, including the tuskless 11-year-old female, whom we called Tuskless. Even though we had not reached the letter *T* yet we made this family the "T" family because of the names already given. Then, because we were not sure how many families were involved or if it was one big family we gave all the other females who associated closely with Torn Ear *T* names.

The handsome female with the first-year calf seemed to associate very closely with Torn Ear, but we were not sure if she belonged to the same family. Harvey named her Tania. We thought she was probably about 30 years old. She appeared to have three sons. The slit-eared female, who we thought was nearing 40, definitely had a very close relationship with the old, straight-tusked female, who looked as if she had to be at least 50. It was difficult to tell if they formed one family or two. We named the slit ear Slit Ear even though it was not a "T" but it seemed so obvious, and we named the old female Teresia.

By the end of 1974, when Harvey had to give up his participation in the project, we had seen these individuals and their calves in various groupings with each other and on their own on 24 occasions. I continued to work part-time on the study for another year, and eventually, with some reservations I decided that the Ts consisted of four family units, and that these four families had close relationships with one another, forming a bond group, which very likely was one extended family. I designated them the TA, TB, TC, and TD families. I also named more individuals in the families: the pretty mother of Right Fang was named Tia; the young female in the TCs was named Tess; a second adult female in the TDs was named Trista; and the two younger females in the TDs were named Tina and Tallulah.

Only two changes occurred in the composition of these four families since the time I had first met them. In March 1974 Torn Ear gave birth to a male calf, the only birth I recorded among the Ts during 1974 and 1975. The second change was the appearance in the TA family of a seven-year-old female calf. The appearance of this calf, whom I named Tonie, remains something of a mystery to this day. I have one picture of her from 1973, when she and individuals I assume were her mother and older sister excitedly greeted the TAs and TBs. I never saw that adult female and the older calf again and I assume they died. Judging by the intensity of the greeting, they must have been closely bonded to the TAs and TBs, possibly even members of one of those families. Whatever the previous history, from late 1975 Tonie remained with the TAs as an integral member of the family.

At the end of the part-time study in 1975 there were 23 animals in all in the four T families. The structure and composition of the families looked like this (calves indented under a female's name were known or thought to be hers):

TA   Torn Ear
      1½-year-old male calf born around March 1974
      4–5-year-old female calf
      8–9-year-old male calf
      Tuskless — a female around 13 years old

Tonie — a 7–8-year-old female (probably the daughter
   of a female who died early in the study)

TB   Tania — the handsome female we photographed in
   1973 with the first-year calf
   2-year-old male calf
   6–7-year-old male calf
   13–14-year-old adolescent male

TC   Slit Ear — the upcurved slit-eared female
   4-year-old female calf
   7–8-year-old female calf

Tess — 11–12-year-old female (probably Slit Ear's
   daughter)

Tia— the pretty, smooth-eared adult female
   Right Fang—4–5-year-old male calf

TD   Teresia — the old, straight-tusked female
   3–4-year-old male calf
   7–8-year-old female calf
   Tina — a young female around 13 years old

Trista — an adult female (probably Teresia's
   daughter)
   5–6-year-old male calf
Tallulah — the young female with the big V out
   of each ear, 11–12 years old

Chapter II

Drought

1976

Slit Ear walked up behind Tia with her ears spread slightly. Tia looked back over her shoulder and quickly moved ahead, leaving the small patch of grass where she was feeding. Slit Ear immediately went to the same patch and fed on it for a few minutes. Tia found another clump of dry yellow grass and began to graze again. She twirled her trunk around the short brittle stems and kicked at the base with her front right foot. Dust and grass chaff flew up as a clump was dislodged. Tia beat the clump against her knee, knocking more dirt and dust and dry bits from it. She then placed what remained in her mouth and as she chewed she tackled another small clump with trunk and foot in an unbroken rhythm. Spread out all around her were the other members of her family and Teresia's family. All the adults and older juveniles fed in the same manner: twirling, kicking, beating, and chewing, and the only sounds were the thuds and slaps of their grazing and the steady flapping of their ears against their necks and shoulders.

The 13 elephants fed for another half hour. In that time Slit

43

Ear supplanted Tia twice more and the young female Tess once. There was no protest from them; they just moved on in search of other small patches of grass. They ate with no apparent enthusiasm. All their movements were slow and deliberate in what appeared to be an effort to husband their energy. They were hungry in a dull, lethargic way, but there was little food available and what there was was mostly unpalatable and of little nutritional value. The younger calves, Slit Ear's and Teresia's, born in 1971, were still suckling from their mothers, but the milk was no longer as rich as it had been and there was not enough of it to satisfy them. They fed on vegetation as well, but they were not as adept at handling the short dry grass. Slit Ear's calf got under the chin of her older sister and ate the grass that she loosened or dropped. Teresia's calf stuck close to his mother. The older calves who had recently been weaned — Right Fang and Trista's son — were having the hardest time. They were not getting any extra from their mothers and had to fend for themselves. As each day of the dry season progressed they grew weaker and more tired.

When the two families finished grazing, they rested for a while under the shade of some yellow-barked *Acacia xanthophloea* trees. After another half hour they moved out and crossed an open stretch of bare, dusty ground to a strip of acacias. Here the older animals broke off small branches from the trees and, carefully maneuvering them with foot and trunk, fed on the twigs, bark, and four-inch-long thorns. The adults were experienced at handling this potentially dangerous food, but the younger calves could neither break off the branches nor manipulate the thorns so that they would not pierce their tongues, gums, and insides of their mouths. They stood near their mothers and picked up the smallest twigs and bits of bark.

After browsing on the acacias for nearly an hour, the elephants walked on west through the strip of trees to an area of the swamp where it runs like a river through steep banks, bordered by *Cynodon dactylon* grass. However, the grass, usually a major item in the elephants' diet, was eaten down to bare stubble and the whole area was covered in the dung of the animals that had fed there.

Slit Ear led the two families to a place where the bank was less steep. Each elephant lowered its front legs over the edge, then kneeling on its hind legs, worked its way down the bank and into the water. The water was only a few feet deep and they were able to cross with no trouble. At the opposite bank they used the reverse technique, scrabbling up to the top of the bank, then kneeling on the lip with their front legs and pulling themselves up over the edge. Teresia and her calf were the last to cross. The calf made it down and across the water but had difficulty getting up the opposite bank. When Teresia walked on to join the others, the calf gave a low rumble of distress, and Teresia, Tina, and Tallulah spun around and came back over to the bank with their ears spread forward and trunks out. They reached for him as he struggled over the lip, rumbling to him. When he made it up, he shook his head and let out a long, loud rumble and the females answered with a softer sound, touched him with their trunks, and then they all turned and walked on to the others.

They were now on an island of hard ground, covered again in eaten-down stubble. Slit Ear did not attempt to graze here but crossed to an area of slower-moving water with thick papyrus and other swamp vegetation. The adults and older juveniles lowered themselves in and began to feed. The younger calves joined them, but the vegetation was coarse and unpalatable. They ate whatever they could find.

No sooner had they begun to feed when they heard the sound of bells of Maasai cattle and the whistles and shouts of their herders. The elephants lifted their heads, raised their trunks, and immediately turned and began to run out of the swamp. No trumpets or rumbles could be heard, just the grunts of the animals as they struggled out of the mud. They raced across the island, slid down the bank into the water, and hurried through, their momentum carrying them up the other side. They ran, with ears laid flat against their shoulders and tails held up and out, back through the acacia strip, across the open dusty plain, and into an area of mixed acacia and palm trees. They finally slowed their pace when they were about a mile from the swamp. They stopped, formed a defensive circle, flapped their ears, and raised their trunks once more, smelling in the direction from which they had come. It was nearing midday and temperatures were already high and the running had heated their bodies. Trista put her trunk into her mouth and with a gurgling sound pulled her trunk back out and sprayed behind her ears with liquid. The others followed her example, some reaching several times into their mouths, sucking water from their stomachs, and spraying their heads, chests, and necks. They then stood flapping their ears rhythmically, cooling the blood in the surface veins of the ears.

Teresia turned and began walking in a northeasterly direction, letting Slit Ear pass her and take the lead. They arrived at the edge of the woodland where it bordered an open, dusty pan. Here they rested under some trees, but after a short time the sound of cattle bells could be heard to the south and the elephants had to move again. They crossed the pan with Slit Ear in the lead, and Teresia in her usual place in the rear. They did not panic this time because the sounds had been from some distance, but they walked at a fair pace. They passed to the north of one of the tourist lodges, ignoring it completely, and eventually arrived at the northern end of the Longinye swamp, where there was no open water. Here there were many other groups of elephants spread out feeding on dry grass. To the south, where the grass was slightly greener from underground dampness, were several thousand Maasai cattle. The elephants, along with buffaloes, zebras, wildebeests, waterbucks, impalas, and Grant's and Thomson's gazelles, were concentrated in a small area where the Maasai did not bring their livestock.

The area still had grass but it was of poor quality. The buffaloes looked like skeletons with old fur rugs thrown over them. There were no young among them at all. The adults had long strings of mucus hanging from their noses, and some had blood-flecked foam on their mouths. The zebras also appeared to be suffering. Many had diarrhea and some were painfully thin, such a rare thing to see in the usually butterball-fat little horses. The wildebeests were also thin and they all seemed lethargic. The gazelles, feeding on small herbs and shrubs, were faring better, but their coats looked rough and dull, indicating a poor diet. The shoulder bones and hipbones of the elephants jutted out and their skin hung loosely in folds and deep wrinkles. Dotted among the living animals were the rotting carcasses and skeletons of those that had succumbed to the drought.

When Slit Ear and Teresia and the others had grazed for a

while they started heading south toward the open water, slowly moving and feeding at the same time. They were still several hundred yards away from the water when a mixed herd of wildebeests and zebras were just approaching the drinking area. The wildebeests and zebras moved cautiously, stopping, looking around, stamping their feet, then moving forward several more paces. About 30 yards up from the drinking place, hidden in the swamp vegetation, were two lions, a male and female. To the wildebeests they were gray blobs amongst more or less gray vegetation. To a human being or another primate the heads of the lions were distinctly tawny against the dark-green reeds. The two lions waited patiently with their ears flattened and their bodies hugging the ground. They had already been there for half an hour watching the herd slowly and cautiously approaching.

When the herd of wildebeests and zebras were less than ten yards from the water's edge they stopped again and remained tense and alert. Suddenly a black-headed heron flew up from the swamp and the herd wheeled and ran back 20 yards. They stood there for another five minutes, stamping and fidgeting, before beginning a new approach. They arrived to within 15 feet of the water and stopped again, but now their thirst took over and the ones in the back began pushing forward, forcing the leaders to move to the water's edge. The first animals entered the water and began to drink, but still the lions remains crouched and hidden. More animals waded in and spread out around the original drinkers until 20 or 30 wildebeests and zebras were drinking in a tight mass. Then the lions made their move. They sat up, which caused instant pandemonium, hesitated a few seconds, and then made their sprint while the animals were turning and running and jostling each other. The two lions ran into this maelstrom, and the lioness, picking out one young wildebeest, closed in, grabbed it by the head, and tossed it to the ground, quickly getting a hold over its mouth and nose. Before the dust had even settled the male joined her, holding on to the rear quarters. The lioness kept a firm hold on the wildebeest's mouth, preventing it from breathing. There was barely a struggle, but even weakened by the drought the animal took several minutes to die. In the meantime, coming out of various hiding places along the edge of the swamp were several more lions: another female and two nearly grown cubs. They began to feed before the female relinquished her hold on the wildebeest's nose. Finally, after nearly ten minutes, she released her grip. She lay panting from her effort and did not begin to feed at once.

By now Slit Ear and Teresia were about 50 yards away, coming down to the water's edge for a drink. Lions cannot kill an adult elephant, but they can kill a baby, and in general elephants are intolerant of lions. Teresia in particular was antagonistic toward lions, probably because of an unpleasant experience sometime in her long life. The elephants had seen the running wildebeests and zebras at the edge of their vision but had not yet seen or smelled the lions. When they got closer, the wind shifted slightly and they picked up the lions' scent. Under certain circumstances the elephants might have just altered their path and drunk at another spot, but their choice of drinking places was limited by the Maasai, and in any case Teresia was not in the habit of letting lions change her plans. She came forward from her usual position at the rear and walked directly and quickly toward the lions, with her head held high and ears spread. The lions were busy feeding, growling and slurping and in general making a considerable amount of noise. Teresia was almost upon them before they noticed her. They took one look and scattered in every direction, and then skulked into the reeds to hide. Teresia swung her trunk at them as they ran away and blew down through her nose. Her family came up behind her. Less self-possessed than she in the face of five lions, they, especially the younger females, were clearly excited. They flapped their ears, gave sharp shakes of their heads, and let out shrill trumpets. They all milled about, rumbling and greeting and reassuring one another with trunk touches. After a few minutes they calmed down and went on to the water's edge to drink.

Slit Ear and Teresia continued south in search of slightly better vegetation, but always keeping the Maasai and their cattle at a considerable distance. The rains were not expected for several more months and the food was rapidly disappearing. The lion incident was a rare distraction in an otherwise relentless pursuit for food. Teresia was bringing her 50-plus years of experience into play to try to see her family through yet another drought, but some of the places she had gone to when she was younger were closed off now and there were fewer choices in the face of this drought.

The lions returned to their kill and squabbled again amongst themselves for the best pieces, but in a mild sort of way. For the predators there was a superabundance of food, and their next meal was waiting just over the rise.

Droughts are not unusual in Amboseli. The name Amboseli comes from the Maa word *empusel*, which can mean "salty dust," "open plain," or "barren place," depending on one's source. It is an area of unpredictable climate and changing environment. By its annual rainfall alone Amboseli would be considered a near desert. It receives only an average of 12 to 13 inches of rainfall per year. The area's rains tend to bypass Amboseli because of a combination of wind and temperature currents created by Kilimanjaro. The mountain, less than 25 miles away, gets most of the rain.

However, Amboseli is by no means a desert. At the same time that Kilimanjaro prevents rainfall from reaching Amboseli, it provides water by other means. Underground aquifers run from the mountain with fresh, clear, cold water which has percolated down through the porous lava from the very summit. The aquifers run for about 25 miles to bubble up in springs in the Amboseli basin. There the water forms channels, pools, swamps, and marshes, and eventually flows and peters out onto the plains and into Lake Amboseli.

The name *Lake Amboseli* has probably enticed and then disappointed many a traveler hoping to find a lakeside camping site cooled by fresh-water breezes. The lake is a cracked, bone-dry pan for most of the year and floods only seasonally for short periods during the rains and some years not at all. Present-day Lake Amboseli is the remnant of a much larger Pleistocene lake that occupied an extensive area right up to the lower slopes of the mountain. The area that this former lake once covered, approximately 230 square miles, is commonly referred to as the Amboseli basin. Within the basin lie the 150 square miles that comprise Amboseli National Park.

Geological evidence shows that during the late Tertiary, a large river flowed southeast through the park, carried on down through Tanzania, and emptied into the Indian Ocean at present-day Tanga. In the late Pliocene or early Pleistocene, 2 to 4 million years ago, Kilimanjaro emerged as a result of volcanic activity and its bulk and surrounding lava fields dammed the river, which then had no outlet. Lake Amboseli was formed. During much of the Pleistocene, the river and runoff from the surrounding higher ground deposited sediments in the lake. Volcanic ash was added by periodic eruptions on Kilimanjaro. Eventually the lake filled with sediments and the river was redirected. After several cycles of drying out and flooding, the majority of the lake became desiccated, leaving only a portion to the west that floods periodically. This final desiccation period may have taken place only 5,000 to 10,000 years ago, which is very recent in geological time scales.

The fact that the Amboseli basin was a former closed drainage lake has had, and continues to have, a profound effect on the ecology of the area. For thousands of years salts washed from the soils were carried down the river and deposited in the lake. When the lake dried up these salts were concentrated in the soil. Under these conditions only a few salt-tolerant plant species could grow. Eventually, with rainfall, the salts leached down through the soil in some areas, leaving the surface soil less saline. Other plants could then take hold and even trees began to grow. The salts, however, remained in the water table, several feet below the surface, and would continue to influence the vegetation in years to come.

Amboseli's complex geological history and its current hydrology and rainfall patterns have resulted in an area with a surprising diversity of habitats and fauna. Within the basin there are open plains, bushed-grassland, bushlands, woodlands, and swamps. A great variety of insects, birds, and mammals are found there. Some of the species live in all the habitats, others in particular ones, still others in only one. The large mammal community is both abundant and diverse. There are elephants, rhinos, hippos, giraffes, buffaloes, zebras, 13 species of antelopes ranging from the tiny dik-dik to the 1,200-pound eland, four kinds of primates, three large cats, wild dogs, two species of hyenas, three species of jackals, several species of smaller cats, plus mongooses, genets, squirrels, hares, and odd species such as the aardwolf, zorilla, ant bear, tree hyrax, honey badger, and porcupine. Over 400 species of birds have been recorded in Amboseli.

Man and his livestock have been an important element in the Amboseli ecosystem for several thousand years. The present inhabitants of the area, the Maasai people, have been there for about 400 years. They apparently displaced another tribe of pastoralists whom the Maasai called the *Iloogalala* — "people of the hard teeth" — and who were forced south. The Maasai were traditional pastoralists noted for their fierce warriors, who raided far and wide for cattle. These were considered to be rightly theirs because their god, Lengai, had given them all the cattle in the world. Their military prowess was well known and they were feared by other tribes and by ivory traders, slave dealers, missionaries, and explorers. While Europeans had explored areas of the interior as early as 1848, few ventured into Maasailand. In 1883 Joseph Thomson (for whom the Thomson's gazelle is named) finally traveled extensively in the area, apparently while most of the warriors were away on a long-distance raid.

On that trip in 1883 Thomson passed through what is now called Amboseli. He found a flat dusty area with no trees but spectacular views of the mountains all around. In his book *Through Masailand,* published in 1887, he noted, "In spite of the desolate and barren aspect of the country, game is seen in marvelous abundance."

In an indirect way the Maasai were responsible for this abundance of animals. They themselves did not hunt wild animals for meat or trophies, and they were intolerant of other tribes hunting in their area. The Maasai warriors sometimes killed large dangerous animals such as lions, elephants, buffaloes, and rhinos to prove their bravery, but this had little impact on population numbers. In times of severe drought or cattle disease, they killed antelopes for meat, but they considered them God's cattle, to be used only as a last resort. This attitude meant that the Maasai have lived more or less in harmony with wildlife wherever they have roamed. Throughout East Africa the areas noted for wildlife today are those that are held by the Maasai or other pastoralists.

Amboseli was recognized very early on as a special wildlife area when, in 1899, it was included in the Ukamba Game Reserve, created by the British administration. In 1906 the name was changed to the Southern Game Reserve, which covered 10,700 square miles. This area was a reserve mostly only in name, but the Game Department did try to control poaching and to regulate legal hunting in the area. In the 1930s stricter controls were brought into action, and Amboseli became popular with hunters, photographers, and filmmakers because its wildlife had the reputation of being easy both to photograph and to kill.

In 1948 the reserve was reduced to 1,260 square miles, named the Amboseli Game Reserve, and placed under the administration of Kenya's National Parks Trustees, a parastatal body. In 1961 the same area became a County Council Game Reserve administered by the Kajiado County Council. (To avoid confusion, it should be explained that in Kenya there are two types of protected areas: national parks and reserves. National parks are administered by the central government, while reserves are administered by the local county council with technical advice from the Wildlife Department. Parks are set aside for the exclusive use of wildlife and tourism, while reserves allow limited human activity by local people, such as cattle grazing and timber felling.)

Under county council administration, revenue from a reserve theoretically goes to the people living in the area in the form of development projects initiated by the council. In 1964, recognizing the value of the Amboseli Reserve in terms of income from tourism, the Kajiado County Council set aside 30 square miles of the central part of the reserve for wildlife and tourist activities only, and the Maasai agreed to keep their cattle out of that area. This situation continued for another seven years, when, under pressure from conservationists, President Jomo Kenyatta (Kenya became independent of Great Britain in December 1963) issued a decree declaring that an area of 150 square miles be set aside exclusively for wildlife and tourism. In 1972 the boundaries were demarcated and the area was gazetted as government land. Finally, in October 1974, Amboseli National Park was established and once again came under the control of the National Parks Trustees. The Kajiado County Council retained an enclave within the park where the tourist lodges were situated and still received revenue from both the park and the lodges.

The decision to make Amboseli a park was a controversial one, for the basin swamps were an important dry-season refuge for the Maasai and their cattle. The Maasai, who had always shared their whole range with wildlife, were asked to abandon these swamps, restrict their movements, and still allow the wildlife to move onto their remaining land. Coincidentally, as part of a scheme for the whole of Maasailand, the land outside the park was adjudicated and parceled out into large group ranches, more or less along clan lines, each run by a committee of elders. Negotiations were carried out with these new ranch committees and various promises were made, including alternative watering points outside the park, grazing compensation fees for the wildlife that came onto the ranches, plus the provision of a medical dispensary, school, and community center. During the upheaval of negotiations, relationships between the Maasai and the government became tense and one of the results was deliberate spearing of rhinos and elephants as a form of political protest. Eventually an agreement was made, but the Maasai were not asked to leave the park as soon as it was established. A pipeline taking water out to the ridges to the north and east had yet to be completed.

When I started my study in September 1972, the wildlife-human situation was not at its best. Nevertheless, Amboseli seemed an ideal place to study elephants. Despite the spearings, the Amboseli elephants were one of the last populations of relatively undisturbed elephants in East Africa. In almost all other areas, elephants had been compressed into national parks and reserves through human expansion. For example, between 1929 and 1959, Uganda's human population rose from 3.5 to 5.5 million. In 1929 elephants occupied approximately 70 percent of all the land in that country; in 1959 they occupied only 17 percent, having been forced into the security of a few national parks and forest reserves as human activities expanded into their ranges. In Lake Manyara, where Iain Douglas-Hamilton worked, the elephants had lost 75 percent of their former range in the 50 years preceding the study. At the beginning of his study there were 480 elephants living in approximately 35 square miles, giving Lake Manyara one of the highest elephant densities ever recorded — nearly 14 elephants per square mile.

In almost all the areas where elephants have lost range it has been due to the spread of agriculture. Elephants and man cannot coexist where agriculture is practiced because elephants will eat and trample crops. Elephants and pastoralists can coexist far more easily because they are not in acute conflict, although they may be in some competition for the available grazing. Amboseli's elephants had been lucky in that they shared their range with traditional pastoralists who more or less tolerated them.

In 1972 the elephants were still moving in and out of the central Amboseli basin in the same migratory pattern that they had been using for several hundred years. Although eventually the legal protected area was only 150 square miles, the whole range that the elephants were able to use was about 1,200 square miles. With a population of approximately 600 elephants, that meant a density of only two elephants to the square mile. Even considering the area within the basin where the elephants concentrated, there was only a density of about three elephants per square mile.

Amboseli's elephants were also fortunate and unusual in 1972 in that they were not being systematically and intensively poached for ivory. The early 1970s was the beginning of a period of severe elephant poaching in most of Kenya. The price of ivory had risen sharply and in the space of only a few years elephant populations were decimated throughout most of the country. In Amboseli there was little or no ivory poaching in the park itself, but when the elephants left on their wet-season migrations they encountered armed poachers. Some of these poachers were Maasai, but most were people from other parts of Kenya and even from other countries. The Maasai were not tolerant of these intruders and often reported them to the Wildlife Department. In this way large-scale poaching in the area was kept in check.

Thus Amboseli's elephants, being neither heavily poached nor compressed into a small protected area, provided a unique opportunity to study a population that was relatively natural, that is, one that was responding primarily to environmental pressures rather than to predominantly man-made pressures. I hoped that a study of undisturbed elephants would provide base-line data that might eventually help in the conservation of all elephants.

When Harvey and I started the part-time study in 1972, I did not realize that within a few years I was going to witness some of the most severe environmental pressures that many of the Amboseli elephants would probably undergo in their lifetimes. Rainfall in the first three years of the study was average to lower than average, but at the same time Maasai cattle had increased considerably, so there was a very high biomass of animals competing for the rapidly diminishing resources. By the end of 1974 conditions were poor. Then followed two years of very low rainfall with almost no vegetation production. The wildlife and the Maasai and their cattle suffered greatly. Many wild animals died and the Maasai lost over half their cattle. It was an extremely stressful time for people and animals alike.

In September 1975, after a period of uncertainty about the future of the project, when Harvey changed jobs and gave up working on the elephants, I finally got a grant from the African Wildlife Foundation and began the more intensive phase of the study on my own. I set up a permanent camp in the park and began to live there the majority of the time. For the first four months I used my own small Renault car; in January 1976 AWF turned over to my project an eight-year-old Land-Rover, which made getting around a great deal easier.

Living in the park right among the elephants changed the study considerably. I was now able to gather information on their daily movements, association patterns, and behavior. Some families I saw several times in a week and I began to know them well. Torn Ear's, Tania's, Slit Ear's, and Teresia's were among those families. I began to do special studies on some of the groups, trying to find out if there was a clear-cut dominance rank among the females within a family.

I would find one of the families I was concentrating on and spend several hours with it, either watching one individual for a given amount of time or observing the whole group and noting particular behavior. The TCs and TDs were two of the family units I had chosen. It did not take many hours with Slit Ear to discover that she was a very dominant animal who did not hesitate to push subordinates around. Teresia was also of high dominance status, in fact she was dominant to Slit Ear, but she rarely supplanted other individuals. Simply put, some elephants seemed to be more aggressive than others. All the same I became very fond of both Slit Ear and Teresia, Slit Ear because she had such a regal demeanor and Teresia because she was old and dignified and gentle and seemed very, very wise.

The other animals in the TC and TD families also became well known as the months went by. Tia kept very much in the shadow of Slit Ear, and young Tess was often the object of Slit Ear's aggression. In Teresia's family Trista was a quiet, reserved female, Tina a slender, unobtrusive young female, while Tallulah and Teresia's younger daughter, whom I named Theodora, were at times flamboyant and playful.

Tallulah and Theodora showed this lighter side of their nature only a few times in 1975 and 1976. For the most part they had to concentrate on finding enough food as the drought became more and more acute. In November 1975 the normal "short rains" were expected, and clouds built up and rolled in from the east, tantalizing both man and animals, but when the rains came they were disappointing. In December I gave up hoping for more, and realized that the animals would have a hard time getting through January, February, and March before the "long rains" were due.

Finally, in April 1976, the long rains came, but they proved to be inadequate, just enough to bring up the barest flush of green. With no further moisture the growth soon stopped. In the midst of the drought, several calves were born to members of the T families. On April 7 Tia was found with a tiny newborn baby, another male, making her the mother of two sons; in May Slit Ear had a calf, another daughter; in June the young female, Tuskless, Torn Ear's daughter, gave birth to her first calf, a female; and in July Trista gave birth to a female.

Other females in the population gave birth as well, and in this year of severe drought, it appeared to be very difficult for the mothers to keep these calves alive. In many cases the mothers altered their normal behavior in what seemed to be an attempt to save their calves. From February to August 1976, I recorded 26 females with new calves. It surprised me to find many of these females separated from their families, just moving with their new calf and perhaps an older calf or a young female from the family. As family units are speculated to have evolved primarily for their antipredator benefits, particularly for young calves, it seemed strange that the mothers would leave their families when their calves were so vulnerable.

I soon realized that one of the reasons the females were separating was that they were restricted as to which habitats they could move in with their new babies. One day I watched the FA family come to the Enkongo Narok swamp in the morning. On this day the female Fiona was recorded with a new calf. The rest of the family plunged into the reeds and waded out into water over six feet deep. Fiona stayed close to the shore with her new baby and fed on the sparse vegetation there. Eventually her family moved to the other side of the swamp and got out on the opposite shore. Fiona was left on the other side with just her new calf. I saw the two of them on their own for several days after that.

Over half of the females who gave birth in 1976 separated from their families for periods ranging from a day to several months. Most of the mothers who split were young females living in families that contained older females who were dominant to them. Under the particularly harsh conditions that prevailed, they may have broken away in order to avoid competition with the bigger females.

Nevertheless, leaving the family appeared to be a last-ditch effort on the part of the mothers to save their calves. In the end nearly half of the 1976 calves died. Every day I found mothers with thin, lethargic little calves moving along the edges of the swamp. A few days later I would find the mother back with her family but with no calf. I never found any carcasses of dead calves. I suppose the predators and scavengers devoured them quickly, or they died in the swamps, where their bodies disappeared in the water and mud.

Of the first-year calves that died in 1976 about half died when they were less than two months old, and the rest died when they were about five or six months old. The newborn calves probably died because their mothers did not have adequate milk supplies. The calves may also have been of low birth weight to start with if their mothers were undernourished during pregnancy. The stronger ones were able to survive for several months, but when they would normally begin feeding on vegetation to supplement their milk diet, at between three and four months old, there was almost nothing for them to eat.

I was saddened but not surprised to find the first-year calves dying. What did surprise me was the disappearances of four- to five-year-old calves. Calves this age were found completely separated from their families feeding in the swamps by themselves. I speculated that they were weakened because they had recently been weaned (elephants give birth to calves about every four to five years and a calf suckles until the next calf is born). It appeared that they were no longer able to keep up with their families in the normal daily movement patterns.

I will always remember one bleak day in August 1976 when I found one of my favorite elephants, Amy, with her three calves: her male calf born at the end of March, her daughter, Audrey, who was then four years old, and her son Albert, who was about nine years old. Amy stood with her head hung low, resting while her calves ate bits of old yellow grass around the base of a *Suaeda* bush. Audrey walked over to Amy and rubbed her head against her rear leg. Amy turned, rumbled, and mildly threatened Audrey, who moved away. Then the baby started to approach Amy, and as he tried to pass Audrey, she kicked backward at him in a halfhearted way. He walked around her and began to suckle from Amy. Audrey moved closer and slowly reached backward with her hind leg and gently touched Amy repeatedly. She must have been very hungry. Audrey had been suckling up till the baby was born in March, but was now refused.

In September I found Amy's family, the AA family, near my camp in a highly agitated state. They ran away when I tried to drive near them. The matriarch, Wart Ear, had a fresh spear wound through her ear. The young female, Alison, was streaming with temporal gland secretions and her baby, born the month before, was gone. Amy was missing. I was very upset. I had become caught up in the lives of Amboseli's elephants, especially the well-known ones such as the AAs and the Ts. When I saw Wart Ear with a fresh spear wound my stomach began to churn and knot and my palms began to sweat. I knew then that I was no longer a truly objective observer. I sat there in the Land-Rover telling myself that I must be scientific and impartial but my stomach told me that I hated seeing Wart Ear wounded and I very much did not want Amy to be dead.

Unfortunately, there seemed no end in sight to the drought and the deaths. Conditions were very severe and no rains could be expected until November. I began finding elephant carcasses and getting similar reports from the warden. Each time I located a group of elephants I carried out a census with trepidation, hoping I would not find anyone missing. Cattle were dying by the hundreds, and as a result Maasai tempers were short and they were spearing elephants and rhinos, not for their tusks or horns but apparently simply in frustration. I was angry with the Maasai for spearing elephants but I also sympathized with them. It was pitiful to see the herds of gaunt, bony cattle moving each day through the park in search of grass. I dreaded driving out each morning because I knew I would see the dead and dying cows everywhere. When a Maasai cow became too weak to follow the herd it lay down and refused to get up again. It usually took about two days to die. It would lie there looking peaceful, chewing the cud (made up of what I do not know). The gullies on the sides of the roads seemed to be the most common place for cattle to die. Each evening there were new ones lying alongside the roads as I drove back to camp. I knew there was nothing I could do for them and I just had to drive by. In the morning some would be dead and some would still be lying there waiting to die.

I did not see the AA group again for 20 days after the day I found Wart Ear wounded, and when I did find them Amy was there but she had a spear wound in her side, and her baby was gone. Amy's breasts were shriveled and shrunken and she looked thin and ill. She stood touching the wound with her trunk and from time to time spraying dust on it.

On the next few sightings both Wart Ear and Amy seemed to be recovering, but at the end of October I found the AAs, and after repeated counts trying to make eleven and getting only ten, I saw that Audrey was missing. I thought she must have been speared by the Maasai. Four days later I saw the family again and there was Audrey! She was with her family for several more days and then on November 20 she was gone again. Later that week I found a four- to five-year-old calf completely on its own feeding in the Enkongo Narok swamp. When I went closer I saw that it was Audrey.

I had seen lone calves of her age before staying near the swamps. I did not recognize any of them individually and thus did not know what eventually happened to them but I assumed that they were sick, lost, or abandoned and that they died. I thought it was the beginning of the end for Audrey. Presumably she was weakened by the loss of milk in her diet, combined with the bad conditions. Instead of following her family out to the woodlands at night, she refused or was unable to leave the swamp edge, and stayed there feeding and conserving her strength. I saw her with her family one more time, and then in the swamp again on her own. Then I saw the family several more times without her. I thought that she had died by then.

Finally in December some good rains fell and the drought ended. On January 5, 1977, I found the AAs and was very surprised and happy to see Audrey following right behind Amy like a normal mother-oriented calf. She has stayed with her family ever since. To me, it was an amazing feat of survival. Lions could easily kill a calf her size, and it would seem that natural selection would wipe out calves that tried to use this "loner" strategy. Subsequent evidence has shown that most calves who leave their families die, but somehow Audrey beat the odds.

The drought took a mighty toll on the elephant population. In 1976 a minimum of 48 elephants in the family groups died, and at least 30 independent males died as well. Of the cows and calves, 17 were adult females who were either speared or shot, 14 were first-year calves, and the rest were older calves. Thus I knew of 78 elephants that had died, but as I did not know the whole population in 1976, I estimated that several more animals, who had not yet been photographed and registered, had also died.

The T families got through the drought with no deaths at all, which was unusual. Even the four young babies survived. The Ts lived in the central part of the park, where heavy tourist traffic may have inhibited Maasai spearings; and they spent their daytime hours in the Ol Tukai Orok woodlands and the Longinye swamp, which were probably the best habitats available. In addition the families contained one old and very experienced female (Teresia) and two large mature females (Slit Ear and Torn Ear), one of whom (Slit Ear) did not hesitate to use aggression in order to get enough to eat for herself and her calves. More than anything I would guess that it was Teresia's knowledge which carried them through the drought.

By the end of 1976 I knew the Ts very well and had named most of the remaining youngsters. In general I now named all

individuals who were weaned. Unweaned calves were referred
to by their mother and their year of birth — for example, Torn
Ear's C (for "calf") '74, or Slit Ear's C'76. I named Torn Ear's
older calves Tilly and Teddy, Tania's eight-year-old son Taabu,
which means "trouble" in Swahili, Slit Ear's two older daughters
Tamar and Tara, Teresia's son Tolstoy, and Trista's son Tim.
Tania's older son, who became fully independent sometime in
1976, was given the male code M84 and taken off the family list.
The four T families contained the following 26 animals (the
numbers in parenthesis next to a name are the estimated or
known year of birth for that elephant and the symbols indicate
sex):

TA	Torn Ear (33)	♀
	C'74	♂
	Tilly (70)	♀
	Teddy (66)	♂
	Tuskless (62)	♀
	C'76	♀
	Tonie (67)	♀
TB	Tania (44)	♀
	C'73	♂
	Taabu (69)	♂
TC	Slit Ear (36)	♀
	C'76	♀
	Tamar (71)	♀
	Tara (67)	♀
	Tess (63)	♀
	Tia (50)	♀
	C'76	♂
	Right Fang (70)	♂
TD	Teresia (22)	♀
	Tolstoy (71)	♂
	Theodora (67)	♀
	Tina (62)	♀
	Trista (38)	♀
	C'76	♀
	Tim (69)	♂
	Tallulah (63)	♀

Chapter III

Migration

1977

The four T families, with their matriarchs Torn Ear, Tania, Slit Ear, and Teresia, stood bunched together forming a single tight-knit group. With them were several independent, but young, males. In all they made up a group of 30 animals. Earlier in the evening they had moved out of Longinye swamp up into the long, tough *Sporobolus consimilis* grass just to the north of the swamp. Now they waited in the security of the tall "elephant grass" for darkness. It had rained for several days in a row and there were strong smells of damp earth and new grass on the wind. The elephants did not rest; they milled about, clearly stimulated and on edge. They were as thin as ever from the long drought but their whole demeanor had changed. Instead of being slow and plodding, their gestures were now energetic and lively. There were frequent rumbles from various individuals and a reaching of trunks toward one another. The younger animals in particular seemed eager to get going, but the big females remained stationary.

Finally, when the sky was lit by only the new moon and a few emerging stars, Torn Ear made the soft "let's go" rumble while slapping and sliding her ears against her neck and shoulders, and set off toward the ridges above the basin to the north. They moved away from the protection of the long grass out onto short grass plains, which had been reduced to nearly bare ground by the long drought. Now the plains were muddy and here and there completely flooded by the recent rain. The elephants were nervous and did not vocalize as they traveled, but at times their presence was given away by the slurping sounds of their feet in mud or their splashing through water. They soon reached the red-soil ridge, which was well drained and less muddy. Although small bushes here had recently flushed green, there was no new grass yet and the elephants did not stop to feed. They climbed onto the ridge, crossing the park boundary halfway up, and headed slightly to the northwest toward the long arc of the Ilaingurunyeni Hills. They walked fast, rapidly covering ground, and they eventually came to an area at the base of the hills that had received more and earlier rain than the places they had come through. Here, fresh bright-green grass was growing.

The elephants began to feed immediately, wrapping their trunks around the stalks, breaking off as big a bunch as possible, and stuffing it into their mouths. It was the first sweet, nutritious grass that they had had for many months and they ate as if they were never going to see any again. The young calves born during the drought had never tasted new grass. They had begun feeding on vegetation, trying to supplement their mothers' milk, when they were four months old, but the only grass available to them in those months was *Cynodon dactylon*, which was both relatively unpalatable and low in nutritional quality during the dry season. Now they found themselves in a sea of delicious, highly nutritious perennials — *Sporobolus nervosus* and *Sporobolus ioclados* — grasses that shot up as soon as it rained and were eaten down almost immediately by the grazers.

The four young calves born in 1976 were not sure what to do. They had never been out of the basin in their short lives and they were nervous and excited. They were also tired from their 20-mile walk. Still, the smell of the grass as it was crushed in the huge grinding molars of their older relatives was irresistible, and although not completely adept with their trunks, they too began to feed.

The elephants fed for the rest of the night, not stopping to sleep as they usually did. As they fed they continued to move to the northwest, eventually arriving at a wooded area on the lower slopes of the hills. By daybreak they were well in among the trees. Here they stopped to rest, but they did not give in to sleep in the same way that they did in the park. The young calves lay down as usual, but the adults remained standing and frequently opened their eyes and lifted and spread their ears, listening for any sounds of danger.

They spent the rest of the day alternating between resting and feeding, not moving very far in their foraging. That night they fed again out in the bushed grassland. Around midnight they were joined by a large aggregation of 80 elephants, made up of eight family units and several bulls, including some older males. The aggregation included the biggest bond group in the population, which consisted of the CAs, LAs, MAs, VAs, and WAs, plus another bond group, the CBs and OAs, and the largest single family, the PAs. The Ts readily mingled with these families and immediately became more relaxed in the security of numbers.

While their fear of danger was alleviated, their excitement increased in the new social situation. The big adult females in the T families were more or less blasé about meeting up with the other families, but the younger animals were curious about, and attracted to, the new arrivals. During the two years of low rainfall each family tended to keep to itself, joining its bond-group families only from time to time and other families rarely. As a result, in those years an individual tended to find itself in a small group rarely numbering more than 20. Finding themselves in a group of over 100 was socially very stimulating. The young T males — Teddy, Taabu, Right Fang, Tolstoy, and Tim — were the first members of their families to make contact with the other elephants.

Teddy and Taabu, being the oldest, set off first and approached the VA family. Here they sought out the young males of their own age. They knew this family as they knew all the families in the population, but some they encountered more often than others. The VAs shared the Longinye swamp with the Ts in the dry season, but during these harsh years they rarely moved together and more often they merely passed at a distance. Teddy and Taabu had not sparred with their age-mates in this family for many months, not only because they had not met often but also because they did not have the energy to play. Now, with stomachs full of nutrient-rich grasses, they could test their strength against the other males of their size and age. They soon found Vladimir and Vostok. Taabu, who was younger than Teddy but always more outgoing, approached the two VA males, and when he reached them he put his trunk first to Vostok's mouth and then to Vladimir's. They in turn put their trunks to his mouth. Teddy came up and exchanged greetings as well. Their four heads were close together and they continued to reach their trunks toward each other, touching a tusk here or a temporal gland there or entwining one trunk with another.

Taabu pushed gently on Vostok and at the same time placed his trunk on Vostok's head. Vostok turned to face him head on, lifted his head, and moved forward, locking tusk against tusk, and resting nose against nose. They stood head to head, pushing gently against each other, moving slowly and languidly, with trunks roaming about each other's faces and in and out of each other's mouths. Suddenly Taabu lunged, hitting his nose hard against Vostok's nose. Vostok lunged back and their tusks struck with a resounding clunk. They began their game in earnest. Vostok was the taller but Taabu was more compact and possibly heavier. It was a fairly even match. They pushed and shoved, backed off and rushed together. Vostok pushed and twisted and managed to poke the tip of his tusk into the skin just below Taabu's eye. Taabu backed and twisted away and turned and ran with Vostok following at a run. Taabu ran around a small bush and spun around to face Vostok over the bush. This gave them both a buffer and they appeared to spar even more fiercely, using the bush to keep them from clashing as closely as before.

Soon they backed off, tired from the unaccustomed activity; and with neither wanting to resume right away, they distracted each other by attacking the bush itself and picking up sticks and throwing them at each other or back over their shoulders. After a few minutes they regained their strength and Taabu knelt down in front of Vostok and dug his tusks into the ground in a direct challenge. He stood up and they rushed together again, tusks clunking. They continued to spar for another few minutes, but then Teddy, who had been sparring with Vladimir, a much younger elephant, grew bored and came over to join Taabu and Vostok. He moved in, breaking them up and challenging Vostok, and they were soon engaged in a match.

Taabu turned and, looking around to see what other prospects there were, saw Right Fang sparring with Vasily. Taabu rushed over and poked Vasily in the backside with his tusks. Vasily bellowed in surprise and some pain and his mother, Virginia, one of the biggest females in the whole population, started over with her ears spread. Taabu stood his ground for a few moments, but when Virginia used the unmistakable threat posture of folding each ear to form a horizontal ridge across the middle, he turned and ran back to his family. The other T males, seeing him running, quickly followed.

The T females and younger calves had remained slightly separate while the young males were sparring. By now the two groups had truly merged. Teresia as usual remained calm and dignified and generally ignored the others. Torn Ear also minded her own business. But Slit Ear and Tania used the opportunity to reestablish important relationships and to develop new ones. Slit Ear raised her head as Mariana, the matriarch of the MA family, came slowly feeding toward her. Slit Ear strode toward Mariana with her head out slightly and her ears partly folded. Mariana stopped feeding, stood tall watching Slit Ear approach, but then backed and turned rapidly away. Tania, coming close behind Slit Ear, lunged at and chased Marcia, a younger female from the same family. When Tania returned she rumbled and flapped her ears and the other Ts came up to her and greeted her with answering rumbles and ear flaps. The Ts had reinforced their dominance rank vis-à-vis the MAs.

The rumbles among the members of the T families attracted some of the older males who had been with the big aggregation. Three males now approached: Steve, Lexi, and M127 (sometimes known as Flop Ear). They were not amongst the biggest males in the population but they were in their mid- to late twenties, the age at which males begin to compete seriously for females. Steve and Lexi were both handsome males, tall and slim with long, asymmetrical tusks. M127 had the misfortune of having crushed his ear in a fight, which gave him a flop ear, and his tusks were undistinguished. However, he was a few years older than Steve and Lexi and so had the advantage. He arrived at the Ts first and moved in among the females with his trunk draped over his left tusk in the "casual walk" so as not to alarm them. The big females ignored him when he tested them, but the young females, Tess, Tina, Tallulah, Tonie, Tilly, Tara, and Theodora, already excited by the mix of so many elephants, rumbled and milled about, turning in circles and urinating. M127, along with Steve and Lexi, tested them by smelling their genitals and tasting their urine. None of the females was in estrus, but the social stimulation began its subtle work on their hormones.

The merged group of approximately 110 elephants calmed down after about an hour and all began to feed again in earnest. As had the Ts the night before, they fed all night. Just before dawn they arrived at a small stream that ran down from the hills and flowed only in the wet season. Here they drank the rushing brown water, which had a completely different taste from the clear water in the basin springs. At dawn a dark solid mass of clouds approached from the east and soon it was raining with a steadiness and lack of wind and bluster that indicated a long healthy drenching of the soils. The elephants moved off into an area of mixed *Acacia mellifera* and *Acacia nubica* trees, where they bunched into smaller groups and rested with their backs to the rain.

At midmorning, with the rain still falling steadily, they returned to the bushed grassland and began to feed again. By the late morning the storm had passed, and with the sun out the wet elephants were black against the clear dark-blue sky and the green glistening grass. They had a few more hours in this idyllic setting before the sounds of Maasai cattle bells and the whistles of the herders carried across the wind from the east. The Maasai were a long way off and the elephants did not panic; they just began to move and feed slowly to the southwest.

The elephants headed away from the hills and eventually arrived at the major road coming into the park from the west. As they were about to cross the road, two minibuses went hurtling past. The elephants at the front of the group were slightly alarmed by the speed, but they were accustomed to vehicles and did not consider them a danger. They traveled now in a long column without feeding. Lillian, the old matriarch of the LA family, one of the most consistent leaders in the population, was at the front, taking the big aggregation to a place where she knew there would be good new grasses.

The herd entered an area of low bush interspersed with grass to the west of the Amboseli lake bed. Here they spread out and began to feed. In the distance they heard the low rumbles of other elephants and knew there were other groups in the area and that there was no immediate danger. They fed for the remainder of the afternoon and into the night, stopping to rest for short periods every few hours. After the long drought they had a lot of catching up to do, and although the fresh grass was giving them slight diarrhea, they would soon adjust to it.

The aggregation, which included the Ts, fed in this area for another two days. There were no rivers here but they drank from pools of water that had temporarily formed from the rain. By the third day they were farther to the southwest, nearly to the Tanzanian border. It had not rained in those two days and the ground was no longer muddy. The elephants were spread out, feeding among the low bushes. Far in the distance they heard the sound of a car engine, a Land-Rover, not a tourist minibus. It came slowly from the south, stopping from time to time, then starting up and moving again, gradually coming closer. The elephants became more alert and wary when the Land-Rover was about a mile and a half away, but then it stopped and they did not hear it start again. It was now to the west of them and downwind. Teresia and a few of the other wise old females raised their trunks and smelled in the direction that the sound of the engine had come from, but they could not catch any troublesome scent on the wind.

Nearly an hour passed and the elephants fed peacefully. It was midday and they began to get sleepy. A few small resting subgroups had already formed in the shade of some of the larger trees. The Ts were still feeding on the western edge of the aggregation. Torn Ear and her three-year-old son were at the front of the movement. Torn Ear was just reaching for a small succulent herb that was nestled in amongst the grass when a quick movement to her left caught her eye. She whirled toward the movement and there were two men only 30 yards away. Without hesitation she put her head down and charged toward them. She did not even hear the explosion before the bullet ripped through the light airy bone of her forehead and penetrated deep into her brain. She was dead by the time she fell forward onto her head and tusks and skidded along the ground for several feet from the momentum of her charge. Her son was hit next, first in the shoulder, which made him scream with pain and rage and then through his side into his heart.

The instant Torn Ear had seen the men, even before the shot rang out, the rest of the heard knew there was extreme danger because Torn Ear had uttered an alarm call just as she charged. All the others heard it and knew who made it and acted accordingly. Most of the elephants immediately began to run away from the source of the danger, but Torn Ear's bond group instantly came to her aid. They ran toward her and even when they heard the shots and saw her fall they kept coming. The men turned and began to run but let off one volley of shots, missing most of the elephants but catching Tina in the chest with a shot that went into her right lung. These shots turned the Ts and they too began to run away from the danger. In the meantime, the rest of the herd ran to the north, which is what the poachers had hoped for, and two more men were waiting there. The WA family was in the front, led by their matriarch, Wendy, closely followed by the next oldest female, Willa. Wendy ran straight into the guns, but these men were not as experienced as the others, and it took seven shots in her head and neck and shoulders before Wendy fell and died. Willa behind her veered off and caught a bullet through her tail, severing it in half except for one bit of skin that kept the lower portion from falling off.

The whole aggregation was now tightly bunched and running at full speed. The mothers were literally pushing their babies forward to keep their pace up. They ran to the north and then to the east, skirting around the wet slippery lake bed. The Ts, having been farthest west and delayed by trying to aid Torn Ear, were at the rear and Tina was the very last. Her family knew she was hurt; they could smell the foamy pink blood dripping from her mouth. She managed to keep up until she got to the ridge, but the incline slowed her down and she groaned with the pain. Her mother, Teresia, kept dropping back to run beside her, reaching over and touching her with her trunk, but finally Tina had to slow to a walk. The rest of her immediate family, the TDs, and Slit Ear's family, the TCs, slowed down as well, letting the remainder of the aggregation run on, including the rest of Torn Ear's and Tania's families. The TAs and TBs were totally confused and panicked by the loss of Torn Ear and they just ran blindly within the large aggregation.

Teresia took them to the far side of Meshanani, a small hill up on the ridge above the lake. There was some protection here, and Tina could go no farther. The blood pouring from her mouth was bright red and her sides were heaving for breath. The other elephants crowded around, reaching for her. Her knees started to buckle and she began to go down, but Teresia got on one side of her and Trista on the other and they both leaned in and held her up. Soon, however, she had no strength and she slipped beneath them and fell onto her side. More blood gushed from her mouth and with a shudder she died.

Teresia and Trista became frantic and knelt down and tried to lift her up. They worked their tusks under her back and under her head. At one point they succeeded in lifting her into a sitting position but her body flopped back down. Her family tried everything to rouse her, kicking and tusking her, and Tallulah even went off and collected a trunkful of grass and tried to stuff it into her mouth. Finally Teresia got around behind her again, knelt down, and worked her tusks in under her shoulder and then, straining with all her strength, she began to lift her. When she got to a standing position with the full weight of Tina's head and front quarters on her tusks, there was a sharp cracking sound and Teresia dropped the carcass as her right tusk fell to the ground. She had broken it a few inches from the lip well into the nerve cavity, and a jagged bit of ivory and the bloody pulp was all that remained.

They gave up then but did not leave. They stood around Tina's carcass, touching it gently with their trunks and feet. Because it was rocky and the ground was wet, there was no loose dirt; but they tried to dig into it with their feet and trunks and when they managed to get a little earth up they sprinkled it over the body. Trista, Tia, and some of the others went off and broke branches from the surrounding low bushes and brought them back and placed them on the carcass. They remained very alert to the sounds around them and kept smelling to the west, but they would not leave Tina. By nightfall they had nearly buried her with branches and earth. Then they stood vigil over her for most of the night and only as dawn was approaching did they reluctantly begin to walk away, heading back toward the safety of the park. Teresia was the last to leave. The others had crossed to the ridge and stopped and rumbled gently. Teresia stood facing them with her back to her daughter. She reached behind her and gently felt the carcass with her hind foot repeatedly. The others rumbled again and very slowly, touching the tip of her trunk to her broken tusk, Teresia moved off to join them.

The Amboseli elephants have been traveling in and out of the basin on seasonal migrations for at least as long as the Maasai have lived in the area. Maasai oral history describes their own migratory patterns as well as that of the wildlife with which they share their range. For as far back as they can remember, the Maasai, their livestock (cattle, sheep, and goats), and the migratory herbivores (elephants, buffaloes, zebras, wildebeests, and Grant's and Thomson's gazelles) have concentrated around the permanent swamps in the dry season, relying on the swamps for both food and water. Then, as soon as the rains come, they have migrated out of the basin to the ridges to the north and east, the bushland to the west, and the slopes of Kilimanjaro to the south.

The way that the Maasai and their livestock, the elephants and the other species use the various habitats of the Amboseli ecosystem came under study by a young Ph.D. student in 1967. David Western, known to his friends as Jonah, was fascinated by the dynamic nature of Amboseli and how people and animals functioned in this environment. Jonah's study showed that man and wildlife follow a remarkably similar pattern of movements and habitat use in response to rainfall and subsequent vegetation production. For all but a few species that do not need to drink water, the choice of habitats is restricted by the availability of water. In the dry season the only standing water is found in the swamps in the basin. In the wet season rivers and streams flow and small seasonal ponds fill with water in the extensive bushed grasslands outside the basin. The domestic and wild animals feed on the higher-quality forage on the more productive soils surrounding the basin as long as water is available. When the seasonal water sources dry up they have to return to the basin and settle for the low-quality but abundant vegetation in the swamps and woodlands.

When I began the elephant study in 1972, Jonah explained the general picture of elephant ecology in Amboseli to me. From his counts he estimated that there were between 600 and 700 elephants using the Amboseli ecosystem. His counts of just the basin indicated that 200 to 300 elephants came to the area regularly in the dry season. He noted that when it began to rain the elephants formed aggregations of sometimes over 100 animals and moved about in these large groups for several days before leaving the basin. Jonah told me that there were three distinct trails leading to the wet-season areas. One was so wide and deep that when the cartographers were making a map of the area from 1950 aerial photographs, they mistook the elephant trail for a minor road and marked it as such on the map.

All through the early years of the study the elephants moved in and out of the basin on a seasonal basis. In 1974 Harvey and I put radio collars on the matriarchs of three family units: Bertha, Echo, and Sona. Although Echo turned out to be exceptionally conservative in her movements, a real homebody, the radio-tracking data confirmed Jonah's general description of elephant ranging patterns. The data also substantiated some of my own observations on the distribution of individual families in the basin in the dry season.

From the beginning of the study, each time an elephant group was sighted particular data were always recorded. These basic data, which were eventually transferred to a computer coding sheet, were:

Date
Time
Location (1 × 1 km. grid square)

    Habitat type
    Activity (of the elephants when first sighted)
    Group size
    Family units present (represented by individuals present)
    Bulls present

I also wrote down the presence of newborn calves, sick or wounded animals, members missing from the family, the occurrence of estrous behavior, and any unusual behavioral or ecological information. When conditions were good, I would attempt to do a complete census of the family or families present.

As these data began to accumulate, it became clear relatively early on that each family in the population tended to stay in a particular part of the park in the dry season and would very rarely or never be seen in other parts. For example, up until the end of 1976, Teresia's family, the TDs, had been sighted 99 times, and in 98 percent of these sightings they were in the central area of the park, in either Longinye swamp or the Ol Tukai Orok woodlands. On only one occasion were they seen west of the Enkongo Narok swamp, and on only one occasion east of Longinye. The data were similar for the three other T families. In addition the Ts appeared to share this dry-season range with about five other families. On the other hand, the FA family, led by Filippa, spent most of their time west of the Longinye swamp. They were sighted 78 times in the same period — 1972–76 — and in 91 percent of their sightings they were in and around the Enkongo Narok swamp, at the base of the Ilmberisheri hills, or in the Ol Tukai Orok woodlands. They were seen in Longinye and east of Longinye on only seven occasions. They also shared their dry-season area with other families. I decided to designate those families that shared the same dry-season home range as a "clan." There appeared to be four or five clans concentrated around the two major swamps in the park.

However, the elephants that used these swamps made up only part of the population and I wondered what the other elephants were doing. Jonah's aerial counts showed that the 600 or so Amboseli elephants were not all found in the basin at the same time. It was possible that the various families alternated using the basin and the outlying areas, switching back and forth and more or less taking turns, with the result that some were in the basin while others were in the surrounding bushland. However, repeated sightings of the same families in the same areas on at least a weekly basis did not suggest this. As I gradually got to know more and more of the elephants and met and photographed new families, I began to see a division in the population. The families that I met near the central swamps were very tolerant of vehicles and would allow me to photograph them from a few yards away. Then from time to time I would meet a group in the park that ran off when my vehicle was still 200 yards away. With patience and slow, careful driving, I eventually got close enough to photograph these shy elephants. Often in those years I would photograph one of these groups and then not see them again for two or three months.

I got a much better idea of what these infrequent visitors were doing when we put a radio collar on Sona, the matriarch of a small family which, when seen, was found in the western part of the park. From a small plane, Harvey and I were able to plot Sona's movements over several months. We found that she spent most of her time to the south of the park in the lava and bushland at the base of Kilimanjaro or in the bushed grasslands to the west. In the wet season she sometimes went north. Her family was usually found in close association with two other families, Jezebel's and Yolanda's, and I designated these three families a bond group. They came into the park and to the swamps in the dry season, but they appeared to visit for only a few days before traveling back out to the bushland. They probably used some small swamps at the base of Kitirua and the Ilmberisheri hills in the western part of the park for drinking, or they may have come into Enkongo Narok swamp at night to drink. In either case, they spent most of the daytime hours outside the park, in contrast to the elephants that concentrated around the main swamps.

Eventually I discovered 20 family units that appeared to be peripheral to the main swamps. I began to call them the "peripheral subpopulation" and the elephants I saw on a much more regular basis the "central subpopulation." Almost all the peripheral elephants were from the western part of the ecosystem, but in 1976 and 1977 I recorded five families from the eastern side. These eastern elephants were very wary of vehicles when I first began to see them and ran away in panic whenever I tried to approach. Over the months, however, they began to calm down, and I was able to get some photographs of them. It turned out that they were actually doing something different from the western elephants. As they began to be seen on a regular basis and they became habituated to vehicles, I realized that they were attempting to immigrate into the central population. My theory is that they made up a clan that concentrated on the two swamps to the east, Namalog and Kimana, which happened to be out of the park. Some of the more modern Maasai began to farm around these swamps, using the water for irrigation of crops. Once they began to grow crops, the Maasai were no longer tolerant of elephants and they began to harass them. Amboseli Park would have been a part of these elephants' range but it was probably an area they did not use frequently in the dry season. With the Maasai displacing them, the elephants were forced to change their dry-season home range. I do not know if the immigrants met with any antagonism from the elephants already resident in Amboseli; but if they did, they must have persisted, because they became residents and today they spend as much time in the park as any of the original groups and are relaxed and tolerant around my vehicle.

At one point during the gradual immigration of the eastern elephants a confusion arose. One of the better-known groups in the central population was the BBs, the Big Tuskless family. We first met and photographed the BBs in 1973 and quickly got to know them well. This family numbered 12 and included two adult tuskless females (Big Tuskless and Bette) and a left one-tusk female (Belinda). These characteristics made them very easy to recognize. Back in 1974, on two of the trips that Harvey made on his own, he recorded the BBs, the "double tusklesses," as we called them then, and took some photographs which were not terribly clear because they were from a distance and the group was nervous. When he came back from these trips he mentioned that he had seen the double tusklesses and I did not think anymore of it. Toward the end of 1975, when I was living in the park, I was in my camp late one afternoon when the double tusklesses entered the glade at the north end of the camp. They came in, took one look at the tents, and fled in terror. I managed to count that there were 12 animals in the group, including two tusklesses and a one-tusker. I thought their behavior was a bit odd, but dutifully recorded the sighting of the BBs in Q-17 (the grid square for the camp).

In January 1976, I was out in the eastern side of the park when I came upon the double tusklesses. They ran away from my vehicle when I was over 100 yards away. I started to get suspicious. This was not like Big Tuskless — she was a huge, placid old female who barely gave my car a glance when I drove up to her family. I managed to approach the group from another angle and then turn my engine off and watch them with my binoculars. On this day there were 11 animals, including a big tuskless female, a smaller tuskless female, and, wait a minute, a right one-tusk female. I hastily consulted my photo files and, yes, BB's Belinda was definitely a left one-tusker. Who was this, then? I looked more closely at the big tuskless. She had a hole in her left ear, as did Big Tuskless, but it was in a slightly different place. And besides, Big Tuskless had the biggest ears of any elephant I had ever known, so huge that she looked like a caricature of an elephant. This elephant had normal-sized ears. I looked at Big Tuskless's photo. The elephant in front of me was certainly not she. I took some photographs from a distance and returned to camp shaking my head.

In the next month I saw these mirror-image double tusklesses several more times. They were still frightened of my vehicle but I managed to get some more photos of them. They were definitely an entirely different family, which I eventually called the GBs. I named the two tusklesses Gloria and Gladys and the right one-tusker Grace. A couple of years later, when I was going over the old photographs, I discovered that Harvey had taken pictures of Gloria and Gladys on two occasions in 1974 without realizing that they were not our original double tusklesses.

Toward the end of 1976 I felt that I had a fairly good idea of the numbers and distribution of the Amboseli elephants. There appeared to be 48 family units, of which 28 concentrated around the central swamps, five were immigrating in from the east, and 15 used the western and southwestern bushlands. All these families were made up of approximately 350 cows and calves. I estimated that there were about 150 adult bulls and thus the population totaled about 500, which, as a result of the drought, spearings, and poaching, was at least 100 fewer elephants than when I started the study. The period of losses was not quite over yet.

In December 1976, when the long drought finally ended and the elephants eagerly left the basin to find fresh pastures, the poachers were waiting for them. Although Amboseli was not an easy place for poachers to operate because of the Maasai, there were some unscrupulous individuals who cooperated with the ivory hunters. That December at least ten elephants were killed.

When the water holes began to dry up in early January and the elephants started coming back into the park, once again I censused the families with a sense of dread. It took several sightings of the T families before I would accept that there had been mortalities.

The first members of the T bond group that I saw were the TCs, on January 3, 1977. Slit Ear and all the members of that family were there, but Teresia and her family were not with them, which was unusual. I saw the TC family again in a large aggregation in Longinye swamp on January 8. This time Tania and her two calves were there, but again neither Teresia nor Torn Ear. Then, on the sixteenth of January, I saw a little group of five young animals looking nervous and indecisive. The group consisted of Tuskless, Tonie, Teddy, Tilly, and Tuskless's six-month-old calf. Although only 14 years old, Tuskless was clearly trying to act as a matriarch and the others were keyed in to her movements and activities. They were in front of the Amboseli Lodge at a place where vegetables and fruit were set out to attract the elephants, much to my distress and disapproval. Up until this time only a few elephants visited this feeding station. The TA and TB families had taken readily to the handouts as they passed the lodge on their normal route from the woodlands to Longinye. Another female and her offspring were consistent visitors and seemed to spend almost all their time at the various refuse pits around the lodges and park headquarters. Harvey had named the mother Greta Garbage. She disappeared in 1974, leaving the garbage field open to whomever wanted to exploit it.

Tuskless and her little band were now feeding on the cabbage leaves and pineapple husks. It was a place where they probably felt safe and were rewarded with highly tasty food. It was fitting that I should find them there when I first suspected that Torn Ear was dead, because with the loss of Torn Ear, their visits to the lodge were to become a very familiar pattern in the future and eventually a serious problem.

I saw the little group of five eight more times in January and knew by the end of the month that Torn Ear and her calf were dead. This was a sad realization for me because Torn Ear was one of my best-known matriarchs, and I had grown fond of her quiet dignity. It was also distressing to watch the remnants of her family trying to make it on their own.

On February 7, I finally found Teresia and her family back with Slit Ear in the middle of Longinye. I noticed immediately that Tina was missing and that Teresia had broken her right tusk.

In the few months after the deaths, there seemed to be a disturbance in the normal social relationships among the four T families. Whereas in the past they were frequently together, in January, February, and March each family was often found on its own. Tuskless in particular seemed to stay away from the other families and spent most of her time near the lodge. It is possible that without Torn Ear she felt threatened by the older females in the other families. She was, after all, very young, relatively small, and tuskless.

I recorded several other animals missing in January, among them Wendy. I first saw the remnants of that group in the latter part of December. I noted that Willa was in a group with a few other odd animals from the CB and OA families. My field notes read:

1055 A small calf appears near Willa which would be the size of Wendy's C'74 — the top of the right ear folds over part way. Tusks just showing — must be Wendy's.

Willa's tail looks like it's been eaten. The tail is very nearly cut in two about ¼ way up from where the hairs begin. She's swishing it around a lot. She keeps rubbing it against bushes. I wonder if it got caught in a snare. It looks like it will fall off — less than ¼ holding it on. I suppose too a spear could have gone through it or a bullet. But what's happened to Wendy? Willa has no milk. An older calf here is probably Wendy's C'71. Wendy and Winnie are both missing.

On January 6 I found Willa again and the lower half of her tail had fallen off. The group was in tall grass and I could not see if Wendy's younger calf was there. I had a fairly good idea of her age. When I first recorded the family in March 1975 I estimated the calf to be about six months old, so she was probably born in September or October 1974. In November 1976 her tusks had grown enough to show beyond the lip, which indicated that she was two years old or just over. I was curious and concerned to see if this calf would live. Willa, the only other adult female in the family, was not lactating, so this calf had been abruptly weaned at a very early age.

On January 18 I saw the WAs again and Wendy's little calf was with Willa. On February 11 she was there again and on this occasion I noted that Willa's tail had healed. On February 15 I found Willa and the calf in a small group of seven with one of their bond-group families, the MAs led by Mariana. I was able to do a good census of the group; I discovered that Winnie was there and was relieved to know that she had not been killed. It was interesting to watch Willa with the young calf. As far as I knew Willa had not yet had a calf, or rather I suspected that she had given birth to a calf in the past but it had not survived. She was now being an excellent mother to Wendy II, as I began to call the calf (I have never used the same name twice before or since, but somehow this calf took on her mother's name — her code initials arc WND). Little Wendy stayed close to Willa, often right under her chin or close in behind her front leg with her head up against Willa's shriveled breast. Willa gently reached around and touched her from time to time.

During the next months I kept a close watch for Wendy II and she was consistently with Willa and apparently doing all right. Interestingly, it was Wendy's older male calf who did not make it. Although he was five to six years old and fully weaned, he simply disappeared and was never recorded again. Perhaps he had been wounded when Wendy was killed and I did not see his injury when I first saw him with Willa. It is more likely that he became separated from the remainder of the family and was poached or speared or even killed by lions. In the next years I was to learn that male calves were more vulnerable than females and were more likely to die both as infants with mothers and as orphans.

When the long rains came in April the elephants again left the basin and migrated out to the ridges to the north and east, the bushland to the west, and the lower slopes of Kilimanjaro to the south. This period was one of definite transition for the elephants, from a time of drought, danger, and death to a time of plenty and relative peace. On the migrations out in April and May 1977 only two females did not return. One was Quelea, the young matriarch of a small family of three; the other was Trista.

I was in Amboseli for most of April and May, but during those two months I did not see the TCs or TDs. I suspected they were out of the park most of that time. On June 1 these two families plus Tuskless and her group came through the camp. When they caught my wind they became very nervous and started hurrying through with trunks up in the air. This was definitely unusual behavior for the Ts, who were probably, along with the AAs, the most habituated females in the whole population. I noticed that Trista and her baby were missing and was immediately suspicious. That afternoon I drove out and found them again and was able to do a complete census. Everyone was there except Trista and her C'76.

I found these families again on June 4. When I approached, they moved off. I noted:

1620 Teresia leads them away to the west. Others follow. Trista and C'76 and Tania *not* here. Even Tuskless is nervous and demonstrates. Trista must have been shot because they're unusually wary of the car.

1627 They have calmed down, but when another Land-Rover approached Tuskless got frightened and started moving away fast although she was only a few feet from my car and Teddy had even touched the car once. They fed all around it and so seem to accept my Land-Rover but are still wary of other vehicles.

On June 7 I found Tuskless and her C'76 in a big group of 130 elephants. With her was another '76 calf, a female about the same size as hers. On June 8 I found Tuskless and the two babies again, plus Tallulah and Tonie. On this day the extra baby was following Tallulah. I realized that it must be Trista's. I wrote in my notes:

> 1705 I wonder if this baby was on its own for a while because it definitely was not with the TDs on the last sightings. Where could it have been? It's already thin and sunken around the backbone. It slowly follows Tallulah.

On June 10 I was driving through the Longinye swamp searching for elephants when I spotted a small group of four. I drove to them and found Tania and her two calves with Tilly, Torn Ear's younger daughter. To my dismay Tania was clearly very ill. My notes read:

> 1732 Tania looks sick — she's standing doing nothing — her skin looks wrinkled and she's hollow above the eyes. She has at least two wounds on her left side — one near her shoulder and the other high up behind her ear. She dusts and feels the wounds.

> 1740 She starts to feed but not vigorously. She appears to have stopped lactating, her breasts are small. There is a third wound on her rump. They are small but pussy. Would guess they're bullet wounds rather than spear. Her two male calves are sparring and playing vigorously nearby.

Just as I was about to leave, Tania walked right up to the side of my Land-Rover to my window and just stood there, lifting her eyelids from time to time and looking in at me. I do not know what she was doing, but I sensed that she was somehow trying to communicate her distress to me and I was very touched and disturbed. There was, of course, nothing that I could do.

It was particularly sad to see Tania wounded and Trista's calf getting weaker and weaker, because for everyone else life was almost literally a huge picnic. There was a superabundance of food and after six months on a high-quality diet the elephants were fat, healthy, and, to use a word that does not seem very elephantine, frisky. The elephants were behaving in entirely new ways and it was a joy to be with them. Each day I went out I would see fat, happy elephants, doing amazingly silly things. I came to call their general play behavior "being silly" because it seemed to be as accurate a term as any.

Just three days before I found Tania, I had been out on the western side of the park in the afternoon and spotted a large group of about 150 elephants along the edge of the swamp. Most of them were feeding and some were resting, but at 4:40 two animals started gently sparring and I was surprised to see that they were females, something I had rarely recorded before. They were Celeste and Calandre, two young females in the CB family. Soon other pairs were sparring, young males and calves and more females.

At five o'clock a tourist vehicle drove up and all the resting animals woke up. Vida immediately charged the vehicle, which sped off, followed by most of the VA group. Because the car did not stick around to be an object of interest, they turned their attention to my Land-Rover and began demonstrating at me. In classic elephant defensive formation they formed a circle with the big females out in front. They rumbled, flapped their ears, shook their heads, and made mock charges at me. I could only laugh at them, it was so obvious that they were not serious.

In the meantime even more animals had begun sparring and chasing each other and generally running around making a great deal of noise. It became like a three-ring circus and I could only write down some of what I saw:

> The ones that are playing are using the vegetation a lot, throwing bits about, running through it.
>
> Willa just chased a young female.
>
> What is interesting is all the adult females playing. There are just as many if not more young females being silly as anyone else. One female has a whole small tree she's tossing around.
>
> Virginia turns toward Lexi, practically kneels then sprays moisture from her nose at him.

Calandre is one of the silliest of all. She charges a minibus by crashing through the bushes.

They are expending an incredible amount of energy.

Even Grace is playing now. It's an amazing sight. Grace comes racing out, frightens herself when she almost runs into me, *really trumpets* and scares everyone into running off. Gloria and Gladys and others ran toward her when she trumpeted, then they go tearing off running and play trumpeting all across the open stretch to the arm of the swamp.

I knew elephants well enough by then to realize that this was not true fear or aggression. When elephants are frightened they are usually silent. Everything about their postures, gestures, and vocalizations on this afternoon indicated high spirits and playfulness. I drove home smiling and feeling lucky that I had witnessed such a scene.

The death of Trista's baby was inevitable, however, and I just wished it would not take so long. The calf was ten and a half months old when Trista was killed and it was impossible for her to survive without her mother's milk. The other females in the family, particularly Tallulah, kept her close to them and did not let her get separated again, but they had no milk, and so in the end the calf grew thinner and weaker, and when I found the family on June 22 she was gone. There have been stories of elephants adopting calves when a mother has been killed, but I think it is probably a rare event. Lactating females will sometimes allow another female's tiny calf (under two months old) to suckle from them but they will not tolerate older calves. An elephant has to produce a great volume of milk for her own calf and under even the best of conditions she would simply not have enough for an orphan without jeopardizing her own calf's life.

The calf of Quelea, the other female who had been killed, also died within a few months of losing her mother. In her case closely bonded females from other families took care of her, as there was only an older brother left in her own family. She was a year old when her mother died, but this was still far too young to survive without milk. Her older brother Quincy, at eight

years old, had to become independent a few years earlier than
he would normally have, but he made the transition and is still
alive today. With the death of Quelea and her calf, the QA
family became defunct and the population now had 47 family
units.

I was very worried about Tania in the days that followed my
sighting of her in Longinye. When I saw her four days later,
on June 14, I was upset to see that she had other wounds that
I had not noticed before, one particularly bad one on her right
hind leg which was oozing pus. She was, however, back with
Tuskless and the others in the TA family and I took this as a
good sign. On June 18 I found her again with Tuskless and
also Tallulah. I noted that she had a wet spot with blood
behind her left ear and that, sadly, while the others were
playing nearby, she stood quietly with her head hanging down.
Tallulah was "being silly," shaking her head and sitting down
in a very ungainly posture.

I think that day must have been the turning point for Tania
because the following day I found her eating vigorously.
Although she was thin she was walking well. Over the next
month she continued to improve. (Tania was lucky. Elephants
have a problem with wounds. Their skin is so thick that often
the wound closes, or nearly closes, on the outside, preventing
proper drainage. The internal wound becomes septic and the
infection spreads throughout the body and kills the elephant.)
Although Tania recovered fully, she did not resume lactating.
Her C'73 was therefore weaned and needed a name. I called
him Toby.

Once Tania recovered, a realignment of bonds seemed to
take place among the four T families. Tania and Tuskless's
families, the TAs and TBs, started to spend most of their time
together, moving in a group of eight. Without Torn Ear's
leadership they turned to the lodge feeding station and the
garbage pits on a daily basis. Slit Ear and Teresia's families, the
TCs and TDs, were not attracted to the garbage and continued
using their home range in their traditional patterns. These two
families, making up 13 animals, were almost always found
together in one of the most tightly knit of bond-group relationships. At the same time they began to spend less time with
Tuskless and Tania, probably because Torn Ear, who was Slit

Ear's contemporary, was no longer there and because the TAs and TBs had changed their daily movement patterns.

The group structure of the Ts, reduced to 21 animals, looked like this at the end of 1977:

TA	Tuskless (62)	♀
	C'76	♀
	Teddy (66)	♂
	Tonie (67)	♀
	Tilly (70)	♀
TA	Tania (44)	♀
	Toby (73)	♂
	Taabu (69)	♂
TC	Slit Ear (36)	♀
	C'76	♀
	Tamar (71)	♀
	Tara (67)	♀
	Tess (63)	♀
	Tia (50)	♀
	C'76	♂
	Right Fang (70)	♂
TD	Teresia (22)	♀
	Tolstoy (71)	♂
	Theodora (67)	♀
	Tallulah (63)	♀
	Tim (69)	♂

Chapter IV
Mating

1978

Saibulu and Chris, medium-sized bulls in their mid-twenties, raced at full speed in pursuit of Tia. She wove in and around tree stumps and bushes and finally reached open ground, where she took off running as fast as she could. The two males followed close behind, but they did not seem to be as fast as she was or perhaps as determined, and she began to draw away from them. Tia made a large arc and by the time she reached its apex she was 30 yards ahead of the males. She slowed down but continued walking fast and at the completion of the arc she arrived at the point where Slit Ear, Teresia, and the others had reached. The males, standing out on the plain, had given up and were temporarily resting from their sprint.

When Tia returned to her family, the others rumbled at her arrival. Her two-year-calf rushed up to her and opened his mouth and let out a hoarse bellow. She touched his mouth with her trunk and he moved closer and began to suckle. Tallulah seemed particularly excited and came over and reached her trunk toward Tia and smelled her and rumbled and then shook her head.

Ten minutes later another medium-sized bull, Mac, approached the family and began to test the females. When he attempted to test Tia she moved off at a fast walk, with her head held up, her ears lifted and slightly spread, looking back over her shoulder at him. He immediately abandoned the other females he had not yet tested and started after Tia. He released his penis from its sheath and began to get an erection. Tia walked out from the family and Mac followed at the same speed, keeping about 20 yards behind her. When they were about 150 yards away from the others, Tia stopped, and, still at a distance of 20 yards, Mac stopped too. Tia stood for a while picking the dusty soil up in her trunk and spraying it onto her head and back. She was standing with her back toward Mac but at a slight angle so that she was able to watch him. He took a few steps toward her and immediately she began to walk away. They continued on around in a circle until she came back to her group.

By now the other two males, Saibulu and Chris, had returned to the TCs and TDs and two younger males had joined them. Tallulah came out to meet these young males and briefly engaged one in a sparring match. Afterward she rubbed her head against Chris's hindquarters and then stood among the five males. She seemed agitated and alert. The males, however, were not interested in her. When Tia had been back in her group for only a few minutes she was chased by all three of the medium males. She ran flat out again and managed to get away.

The sequence of either running or walking after Tia occurred many more times in the course of the morning, but none of these males ever got close enough to touch her. Sometimes her calf ran with her when she was chased, desperately trying to keep up. Eventually he learned to stay back with the other females. When the group stopped in the shade of an acacia tree, the bulls rested as well. Later in the afternoon the pursuit resumed, and at 4:30 Tia was finally caught by Chris in thick habitat where it was difficult to run. Chris managed to stop her and mount, but Tia kept moving sideways and he was unable to penetrate. At the same time Tia's family came running over and, surrounding the pair, they rumbled and trumpeted. Chris ejaculated all over the back of Tia's legs and dismounted.

Tia stood with her head low, looking tired after all the running she had done during the day. She had had very little time to feed and was harassed even when she tried to drink. Soon most of her family drifted away to begin feeding again, leaving Tia and Tallulah resting. Chris also stood quietly nearby. Tallulah remained alert and was the first of the females to hear the low rumbling sound almost like an engine far in the distance. When Tallulah rumbled in response, Tia listened and could also hear the distant sound. They waited tensely, listening, and after about ten minutes they could smell the sharp, acrid, unmistakable odor of a musth bull — a large, sexually active male elephant in full rut.

Tia and Tallulah watched the arrival of Bad Bull with great interest. He was a massive animal with two jaggedly broken tusks and a large V notch out of the bottom of his right ear. About 45 years old, he stood at least two feet taller than the medium-sized bulls and his head, particularly his forehead and the space between his tusks, was extraordinarily broad. His temporal glands, one on each side of his face, located midway between the eye and ear, were grotesquely swollen and secreting a copious, viscous fluid. His smell, almost intoxicating to the females, originated from the temporal glands and from the area of his penis and sheath. He was dribbling urine continuously, spurting it out at a very fast rate. His sheath was covered in a greenish white scum and his inner legs were black with wetness.

Even though they had all heard and smelled him coming, his arrival caused a considerable commotion. He walked straight for the females, carrying his head high, tucking his chin in, and waving his ears in a characteristic musth male posture. He did not bother to drape his trunk over his tusk and lower his head to reassure the females. They were both attracted and alarmed. First they rumbled a greeting to him, and as he got closer they milled about, turning and backing, some reaching their trunks toward him, others urinating. Even the young calves were fascinated and came over to smell him. The males who had been surrounding Tia, more or less waiting for her next move, immediately ambled off to the periphery of the group, where they started to feed, "pretending" that they had no interest in Tia, but all the time positioning themselves so that they could keep an eye on her and Bad Bull.

Bad Bull did not test the other females, even though Tallulah kept placing herself right in front of him. Tia had walked off a few paces in the characteristic estrous posture of ears held out tensely, head up and turned slightly, looking back over her shoulder, and he noticed her immediately. He turned toward her and approached. As soon as she had his attention, she began to walk faster, heading away from the others. He too picked up speed and at the same time his penis emerged from its sheath and dropped down, nearly touching the ground.

Tia began to run, but not fast and not with the determined intensity with which she ran when chased by the younger bulls. Bad Bull quickly caught up with her. He reached out with his trunk and placed it along her back. She stopped immediately. He moved his trunk forward to the top of her head until his head and tusks were over her hindquarters. Then, using his head as a lever, he reared up, placing his front feet on her backside but sinking back so that nearly his whole weight was on his hind legs. His penis had hardened, changing in shape from a simple arc to an S-shaped double bend. It curved out from the sheath, then down, and then curved back up at the tip. Fully extended and erect, it measured almost four feet and was endowed with musculature allowing directional control. Bad Bull whipped it about, searching for Tia's vulva, which hung low between her legs with the opening facing the ground, not up under her tail as in most ungulates. Tia backed in closer and spread her legs. The S-shaped curve of the penis now came into play. Bad Bull sunk down even farther and hooked the tip of his penis into the opening and thrust upward, forcing the full length of the penis into the vagina, at the same time lifting the whole vulva and protuberant clitoris at least a foot. Tia stayed still and they held this posture for about 45 seconds, after which Bad Bull dismounted and withdrew his penis with a gush of semen.

Tia stepped forward with legs still spread, opened her mouth, and uttered a very deep, loud sound, more like a bellow than a rumble. Her family came running over, flapping their ears and rumbling, screaming and trumpeting in a great outburst of vocalizations. They extended their trunks toward her, some to her mouth, others to her genitals. Tia turned back to Bad Bull and reached her trunk in the direction of his penis, which, although no longer fully erect, still hung down free of the sheath. She then turned and stood parallel to him and rubbed her head against his shoulder. The members of her family began to calm down but Tia continued to give repeated low, long, pulsated rumbles for the next ten minutes, and occasionally reached her trunk to touch Bad Bull's penis or to smell the sperm on the ground. Tallulah remained close to the pair and still showed some excitement. After a while, she moved in close to Bad Bull and stood on his other side.

Bad Bull stayed with Tia for the rest of the afternoon and all through the night. At the same time Tia made sure she was never more than a few feet from him. With him there she could feed and drink and rest without being harassed by the other bulls. They did not mate again until the next morning and this time Tia did not even run from Bad Bull to start the sequence, but once again her family raced over with excitement and Tia made the long, low post-copulatory rumbles afterward. By morning there were eight other bulls with the TCs and TDs, all being very discreet about their interest in Tia, in fear of reprisals from Bad Bull. Several times in the night a male would try to inch in close to the pair, and Bad Bull would lunge at him, sending him running off with a groaning sound. A few times Tia wandered a couple of yards away while feeding, but as soon as a male tried to approach she moved straight back to Bad Bull.

They stayed in consort throughout the next two days and mated only one more time. By the fourth day Bad Bull began to lose interest in Tia. She tried to maintain the consort relationship with him but he did not guard her with the same enthusiasm and eventually he allowed the other bulls to separate her from him. Soon after, he left the TCs and TDs and strode off in search of other estrous females. Without a protector, Tia was once more harassed and chased by the younger bulls. But once again Tia was determined to get away from them and although they were relentless in their pursuit, she was able to evade all but one of them. Hector, a fairly big bull, managed to run her down and mate with her, but the chances of being impregnated by him were low, as she was now at the end of her estrous period. By the following morning the bulls had lost interest in her and they too left the family. Tia returned to her confused and weary calf and resumed the normal routine of elephant family life.

The very next day, Tallulah, who had been stimulated by the presence of the males and the sexual activity, finally came into estrus herself. She was 15 years old and had never been in estrus before. Although she had carefully watched Tia and Slit Ear (who had been in consort with Bad Bull the week before Tia), when Tallulah actually came into estrus herself she seemed to have forgotten everything she might have learned from her observations.

From the moment the males started to show an interest in her she lost her head completely. She ran off in a panic whenever one got within 20 yards of her. Being young and lighter than the males, she was fast and agile and was able to outrun them, but she soon began to tire because of her total lack of discrimination of which males were serious and which were not. She eventually got caught and mated by an insignificant young male, Ed, whom Slit Ear and Tia would never have allowed to catch them. Tallulah rumbled and moaned and screamed throughout the mating and stood with her mouth wide open and bellowed after Ed dismounted. Her family dutifully came running over and vocalized and demonstrated with her. Predictably this attracted even more males.

Within the next half hour Tallulah and her family were surrounded by 17 adult males. Singly, in pairs, and sometimes in threes and fours, they chased Tallulah and eventually completely separated her from her family. She was chased over and over again, once for over a mile, and she was caught and mounted again but unsuccessfully. The male could not gain intromission and Tallulah managed to wriggle out from under him and run off once again. This behavior continued on and off for the rest of the afternoon and into the night. Fortunately for Tallulah, the males got tired as well, and there were periods when they all, Tallulah included, stood in a group and rested. Finally, around 4:00 A.M., a large adult male in musth, Ajax, found Tallulah, quickly dispersed her suitors, and mated with her. She gave the characteristic deep, pulsating rumble after the mating, but since her family was now several miles away, she got no response from them.

Tallulah stood with Ajax after they had mated and looked as though she were going to go into a consortship with him. She followed him when he took a few steps forward and she even rubbed her head against him. However, after about half an hour, some of the more aggressive bulls moved in closer and Tallulah began to get nervous. Ajax was resting with his eyes closed and two of the males, RBG and M135, slowly snuck up behind him. Instead of sticking close to Ajax's side as a more experienced female would have done, Tallulah panicked and ran off, with RBG and M135 in hot pursuit. Ajax woke up and ran after them. They were out on an open plain and all four elephants were running at a terrific speed, kicking up a great cloud of dust. After about 150 yards, RBG caught up with Tallulah, stopped her, and mounted. Just as he was hooking his penis into her vulva, Ajax arrived at a run, came straight for RBG, rammed him with the broad part of his head, and pushed him off Tallulah. RBG staggered away for a few yards before he gained his footing and spun around, and with head held high and ears spread, he started back toward Ajax. Ajax in turn raised himself to his full height and spread his ears. They both stopped and faced each other. Then Ajax stepped forward, at the same time folding his ears horizontally and forming a ridge across the ear. RBG stood his ground for a few moments before taking a step back and turning and walking off rapidly. Ajax followed for about 20 yards and then turned back toward Tallulah.

In the meantime, M135 had taken advantage of the two males' interaction and was pursuing Tallulah across the plain. Ajax quickly caught up with them and M135 immediately veered off. Tallulah was so agitated and confused that she continued to run away and Ajax gave up and slowed to a walk. Tallulah finally came to a halt 100 yards farther on and stood by herself.

Tallulah was in estrus for five days in all and for most of that time she was chased by males. Several males mated with her, including Ajax again. She started to get the idea of a consort relationship by the third day and began to try to stay with Ajax, but the males seemed aware that they could intimidate her and were usually able to separate her from him. Ajax had to spend most of his time threatening other males who tried to come in close and this almost inevitably led to Tallulah being chased off by yet other males.

On the fifth day of her estrous period Tallulah was no longer of interest to the larger males, and she was chased and harassed by young males in their early twenties. Eventually they too ceased to find her attractive and she was able to wander off in search of her family. She went to the Ol Tukai Orok woodlands and, using her trunk to smell both the wind and the ground, she tried to track her family down. She moved along the well-used elephant paths, snaking her trunk along the ground and smelling as she went. Every once in a while she stopped and uttered a very low, long, drawn-out rumble, and then stood still and listened. When she got to the southern end of the woodlands at the edge of an open plain she suddenly stopped short and intently smelled the ground at her feet. Her temporal glands began to flow until the sides of her face from the gland down to her chin were black from the liquid. She started moving fast now, keeping her trunk to the ground. She rumbled gently every 15 seconds. When she had covered about 500 yards in this way she stopped again, raised her trunk, smelled the wind, and rumbled louder. Off in the long elephant grass on the far side of the plain came an answering rumble. Tallulah took off running, covering another 500 yards until she reached the long grass. The members of her family, also streaming from their temporal glands, had started out to meet her and she raced into the middle of them as Teresia, Theodora, and some of the others reached out their trunks and greeted her with loud rumbles, trumpets, and many ear flaps. Tallulah opened her mouth and gave forth a resounding rumble of reunion.

When the good rains came in 1977 and continued into 1978, along with them came such a profusion of sexual activity that at times it seemed like a bacchanalian debauch. But oddly enough, before the Amboseli study little was known about elephant reproductive behavior in the wild. Very few matings had been witnessed and documented and many people concluded that elephants mated only at night or were very secretive about it in the daytime. During Iain Douglas-Hamilton's four-and-a-half-year study in Lake Manyara he saw elephants mating only four times and saw females who were "apparently" in estrus on ten occasions. This was difficult to understand because there were many births during his study, so the elephants had to be mating. Another scientist working on elephants, Roger Short, offered a substantial reward to 50 rangers in a Uganda park if they could find a female in estrus for him. In over four months of searching they could not find one. In the end he happened upon one himself and watched her behavior for two days, and, interested in the physiological aspects of reproduction, he then shot her and collected her reproductive organs. Short concluded that there were no external signs of estrus and that the only way estrus could be detected was by seeing an elephant being mated.

I started my study with no preconceived ideas about elephant sexual behavior, but I did not expect to observe it frequently. I had seen only one mating while working with Iain in Manyara. During the part-time study in Amboseli, Harvey and I saw several matings at close range on one day in 1974 and we were very excited and pleased to be able to witness them. Over the course of six hours, a young female, Katrineka (named after Harvey's daughter), was chased by 12 males and caught and mated four times by a large male, M2. Aside from the fact that she was chased and mated, we could see that she showed no external signs of estrus, just as Short had reported. Her vulva was not swollen, nor did she have any discharge.

During the rest of the part-time study, on one or two occasions I saw females being pursued by males, but I did not see any other matings. After I moved to Amboseli in September 1975 and began my more intensive observations, I might have expected to see more reproductive activity but I did not. During the whole of 1976 I observed only one mating and saw females showing unusual behavior, which I suspected indicated that they were in estrus, on two occasions. At the time this seemed normal, given the record of other elephant workers. What I did not realize was that I was not seeing estrous behavior in 1975–76 because, as a result of the drought, very few females were coming into estrus. Once the drought broke at the end of 1976 and abundant vegetation became available, the females picked up in condition and started cycling again.

Starting in February 1977 I began to see estrous behavior on a regular basis. By then I had been observing the Amboseli elephants for four and a half years and knew them fairly well. Although I had been watching them during drought years when they had been relatively listless, I could detect odd nuances of behavior that indicated a female was not acting in a normal way. I always recorded unusual behavior whenever I saw it and continued to do so even though there seemed to be a great deal of it in 1977. However, "being silly" was one thing, while acting in a peculiar way only in relation to males was another. Once I became sensitive to estrous behavior, I saw it frequently. In 1977 and 1978 I observed estrous behavior on 80 different occasions, during which I saw 19 matings. But mating was not the key indicator of estrus. Other behavior was more common. Eventually I divided the behavioral components of estrus into five categories: 1) wariness; 2) the estrous walk; 3) the chase; 4) the mounting; and 5) consortship.

With older, experienced females, such as Tia, there was a definite pattern to estrus, which lasted from two to six days, with an average of about four days. On the first day or two the female is wary of males and tries to avoid their approaches. During this period there are frequent estrous walks and chases, and if the female gets mated she and her family vocalize loudly and generally make a big commotion. All this activity attracts more males, which is most likely the purpose. During what might be considered the second stage of estrus the female is found by a large male and mates and goes into consort with him. Consorts last from a few hours up to three days. It is a relatively quiet period, during which the male and female maintain close proximity and the female is usually not chased by other males. A third stage of estrus often occurs after the consortship ends, when the female is again pursued by younger males.

In the case of young, inexperienced females who have never, or only once or twice, been in estrus before, the normal behavioral pattern is disrupted completely. These young females, such as Tallulah, get chased and harassed throughout the estrous period and are often repeatedly mated by a series of bulls of all ages. It would appear that, as with so much else related to elephants, consort behavior is something that is learned with experience.

At the same time that I began to piece together the components of estrous behavior in the females, I was beginning to take more of an interest in the males. I am always being accused of being a female chauvinist in that I supposedly neglect the male elephants. While my study definitely has concentrated on the cows and calves, and while it is true that males were assigned numbers and females were given names, I can explain it all. It goes back to my time with Iain in Manyara. He named the females and numbered the males and I simply continued the tradition. In my case it was difficult enough finding over 300 female names, each with a unique three-letter code, without adding male names as well. As it turned out I named many of the males, but used their numbers on the computer coding sheets. Iain also influenced my way of looking at elephants. He definitely focused on the females and concentrated on their social organization. I too found the intricate relationships among the females and their complex social structure intriguing, and that is what I wanted to study when I did my own research project.

By the end of the part-time study, Harvey and I had photographed about 100 adult males. We considered males adults once they had left their natal family units. Males reach sexual maturity around 10 to 12 years of age and leave their families at anywhere from 10 to 19, with the average age of independence at 14. I continued to photograph males when I started my full-time study and continued to collect data on where individuals were, whether in all-bull groups or with cows and calves, and so forth. I did not spend much time actually watching their behavior, except when they were with females.

In October 1975, a few weeks after I set up my camp in Amboseli, Bob Poole, director of the African Wildlife Foundation, which I was working for and was sponsoring me, came to visit me with his family. Bob had a 19-year-old daughter, Joyce, who was at Smith College (coincidentally my own alma mater) in the United States but who was spending a year in Africa before returning to her studies. She was a zoology major and planned to take courses at the University of Nairobi, hoping to gain some field experience with African wildlife. Bob asked if she could work with me and I gladly agreed. He made arrangements, receiving permission from the Wildlife Department, and Joyce came to stay for three months during 1976. From the beginning I felt that Joyce was a colleague rather than a research assistant and I decided to give her a project to do. I assigned her to the bulls. That was the beginning of a long and very fruitful collaboration which continues to this day. I might add that Joyce never felt that she was given the less interesting sex to study. Quite the contrary.

Part of Joyce's work involved bringing the recognition file for the bulls up to date. She took more photographs and worked with the photographs that I had not yet sorted. I had devised a different filing system for the bulls from that of the cows, using a punched-card method. A friend, Chris Hillman, had used these cards in his study of eland and had suggested it might work well with elephants. Each card had 102 numbered holes running around the entire outer edge. I made a key, assigning various recognition factors to the numbers, such as "large hole in left ear," or "right tusk higher than left." Three of the holes indicated three general size classes of the bulls — young, medium, and large. Each bull's characteristics were assessed and the holes for these characteristics cut through on his card. Thus, out in the field, when I came across a medium-sized bull with a large V out of his right ear and a broken left tusk but I did not know who he was, I could take the stack of cards, run a spike through the hole for "V nick right," and all the cards for bulls with that characteristic would fall out from the bottom of the pack. I could then take those cards that had fallen out, run the spike through the hole for "right broken tusk," and more cards would fall out. If the number of cards that dropped down was still many, I could run the spike through the hole for medium-sized bull. By that time only a few cards would probably fall out and I could just quickly look through them.

This system worked well for the bulls because a bull could be anywhere and with anyone and thus there were no clues as to who he was other than his ears, tusks, and size. The cow pictures were pasted onto plain cards and carried in alphabetical order by their family. If one member of the family could be recognized, that provided a huge clue and the pictures of that family could be taken out of the file to compare with the elephants present. Eventually we put the cows onto punched cards as well, and new researchers used the system when they were first learning the elephants.

Back in 1976 Joyce worked hard on updating the punched-card system for the bulls and soon learned most of the medium and large bulls. (The young bulls who *all* seemed to have perfectly smooth ears and symmetrical tusks took quite a while longer. Between us, plus two other researchers, we completed their cards only a few years ago.) Joyce also went out and watched bulls and found them in new places where I generally did not go because they were not areas that the cows and calves used frequently. I often wondered where the bulls were when they were not with cows and what they were doing.

In Manyara Iain had noted that he would see a large bull in the park with the cows and calves for a couple of months and then he would not see the bull again for perhaps a year. Other bulls seemed to live in the park but not mix with the females very often. He suspected that there might be a sexual cycle in the bulls but did not detect a definite pattern. I too tended to see a big bull for a space of time and then not see him again. I never knew whether the bull had been killed or if he was just away somewhere and would come back.

One day in early March 1976 Joyce and I were out in my Land-Rover when we came across a very large bull, the biggest so far recorded by us in Amboseli. He was with an aggregation of cows, calves, and younger bulls and towered over them all. He had just pushed down a small acacia tree and the EA family was feeding on it with him. Evangeline of the EAs was being wary of males and at one point was pursued by several, but not by this huge old male. I had seen him only twice before, and that was in the previous month. He had been assigned number 103. I assumed he was one of the bulls that left for long periods of time and was currently on a visit to Amboseli. We noticed that he was constantly dribbling urine, as if there were some malfunction of the sphincter muscle that controlled its flow. We knew he was an old bull because he was so large. (Elephants grow throughout their lifetime, and thus the larger an elephant is the older it is.) We thought M103 was so old that he had become incontinent.

We saw M103 again a few days later, still dribbling urine. He was with a large group of over 100 elephants that included the EAs and Evangeline, who was still being chased by other bulls. We decided to call him Zeus because he was the biggest of all the males, but we never used his name; that was the last time we ever saw him.

On the morning of March 26, I found another large bull, M117, dribbling urine. He was with seven other bulls. Later in the morning when I was with a small group of cows and calves a few miles away, he suddenly arrived and tested one of the females who had been acting wary and coy around the males that were already there. The next day I found M117 with another group of cows and calves. He was still dribbling and in my notes I wrote:

> It looks like green pus coming out of the opening to the sheath and penis.

On March 30 Joyce and I were out again together and found the AA family with many bulls, including M117 ("the bull with the dripping penis," as I wrote in my notes). By now his sheath had turned a greenish-white color and gave off a very sharp, pungent odor, which was not unpleasant, but very strong. We were convinced that he had a dreadful disease and started to refer to his condition as "green penis disease." The unfortunate bull himself was called Green Penis. I was disturbed that there appeared to be a disease passing among the bulls, but I really became alarmed later in the morning when Green Penis began to show interest in a female who seemed to be in estrus. We saw that his intentions were clearly sexual when he got an erection and followed her into a small grove of trees. Joyce and I both noticed that about a third of his penis had also turned green. I was sure he was going to give the disease to one of *my* females and envisioned a nasty epidemic sweeping through the whole population.

Joyce saw Green Penis two more times, on April 1 and 2. On April 2 she came back and told me, "I smelled him before I even saw him." Joyce did not see him again before she left Amboseli on April 10 and I did not see him again for the rest of 1976. To my relief, I did not find any signs of a female version of "green penis disease" in the next months.

On June 13, 1976, I discovered another bull with green penis disease. He was big, with a broken right tusk and a large V nick out of the bottom of his right ear. I noticed that his temporal glands were streaming, which seemed unusual for a big male. He was with the VA family. I saw him again on June 30, still with streaming temporal glands, and this time I noted that "This bull is dribbling urine like Zeus." I saw him one more time with females in September and then not again for the rest of the year. He was photographed and assigned number 126.

Four months later, in early January 1977, I found M126 and was relieved to see that he had recovered from the green penis disease and was no longer dribbling. I sighted him three more times in the next three months and each time he was with males and seemed fine. In February I found yet another male, M22, with green penis and streaming temporal glands. He was with females and seemed interested in the adult female Delia, whom I suspected was in estrus. It began to look as if there were some connection between green penis, or GP for short, and sexual activity.

Later in the year I began to see another side of GP — one that directly affected me and was not at all to my liking. My first inkling of it was on April 2, 1977. I found a group of about 50 elephants near the Enkongo Narok swamp. With the cows and calves were three or four good-sized bulls and the biggest was M107, David. I wrote in my notes:

> David deliberately comes over to the car twice, coming close and gently threatening by standing tall. The first time I moved the car he followed. He definitely has GP disease and he already smells. He let the females go by and then came back to threaten me. If I didn't know better I'd say he was guarding "his herd."

I meant by "if I didn't know better" that it was clear from all the previous work that had been done on elephants by Iain, Harvey, and others that there was no such thing as a "herd bull" who stayed with a particular family in a harem system. An individual bull never spent more than a few days with any particular family. I had observed the same pattern, but on that day David was definitely being aggressive, which in itself was unusual for an Amboseli elephant, and he seemed to be trying to keep me from getting close to the females.

On June 2, 1977, I met the big, broken-tusked, V-nick right bull, M126, again. He was with a large herd of over 50 cows and calves and at least 12 bulls. The bulls were trailing along at the rear of the group and among them was one young female. M126, along with my favorite of the big bulls, Cyclops, who was a one-tusker, and Pablo, were rubbing against one another and appeared to be playfully sparring. It was a quiet, peaceful scene, with the elephants slowly walking toward the swamp. I pulled my vehicle up parallel to their movement, turned the engine off, and watched the group go by. As M126 came past he suddenly whirled around and came for me with head down and with no hesitation. He stopped about ten feet from the car, stood tall with his ears spread, and towered over me. There was nothing I could do. It was too late to start the car, for I feared that would provoke him. After a short time, he turned away and walked on to join the others.

I moved my car to follow the group to the swamp edge. As I came around a corner I met M126 and he charged me again. I thought he was just feeling rambunctious, for there had been sparring among the males and a lot of playing among the younger animals. I drove on ahead to meet the group as it arrived. I drove down the track about 500 yards and then off the road, winding down through the trees and bushes until I was in thick regenerating acacias at the edge of the swamp. I sat there for a while watching the females arrive at the water. Suddenly I got a funny feeling in the back of my neck, looked in my rearview mirror, and found it filled with dark-gray wrinkled hide. I spun around in my seat and there was M126 bearing down on me. I started the car and just managed to get through a gap in the bushes in order to turn the car around to get back into the open. M126 was right at my tail and as I wove through the trees he actually tried to head me off.

I got into the clear area and with speed left M126 behind. I then stopped the car and sat there shaking for a while. I decided that I could not let him get away with it. When cows or bulls were aggressive toward Iain in Manyara, where aggression was far more common than in Amboseli, Iain would either stand his ground or actually go on the offensive. I remembered being with Iain on several occasions when a nasty male, whom Iain called Mr. Big, charged and threatened us. Iain would start the car, rev the engine, and charge him, and much to my relief Mr. Big always backed down. I thought, "Cynthia, if you want to study elephants you can't let this elephant intimidate you. You have to show him that you are dominant." M126 was standing about 50 yards away, keeping his eye on me. I turned the car around and headed straight for him with as much speed as I could get up over the rough ground. As I headed toward him he started to come for me with head down, trunk curled in, and ears folded. We got closer and closer and at the last second I veered off. I turned the car around and charged him again, but once more I was the one who veered off. Each time he swung after me and chased me at full speed. I had little doubt that he would carry through his charge. I drove on till he gave up and stopped the car once again, trembling but also feeling a heady rush of adrenaline.

I was angry and upset but also curious about what he had done. I turned back down the road to where he had charged me the second time and found his footprints going straight along the road and the trail of his trunk snaking along the ground over my tire marks as he tracked me. He had left the other elephants and had deliberately followed me for nearly half a mile, first along the road and then along my winding route in and around the trees and bushes to the swamp edge. None of the other elephants had taken this route, which was indirect. I found this behavior totally strange, surprising, and not a little creepy. From that moment on I thought of him as "The Bad Bull," and that became his name.

I refused to give in entirely to my fear and stayed with that herd for another two hours but kept looking over my shoulder. The bush was too thick to know where Bad Bull was at all times. In fact I did not see him for more than an hour after our confrontation and when I did find him he was with the young female who had been with the bulls earlier on. She seemed wary and nervous and I thought she was probably in estrus. In the confusion of the events I had not written down whether or not Bad Bull had green penis but I remembered that the sides of his face had been covered with thick viscous secretions from his temporal glands, which is almost always associated with GP.

Joyce came back for several weeks during her summer vacation in 1977 and started to fill in some important parts of the bull puzzle. We were both beginning to wonder whether green penis was really a disease, but we were confused about its occurrence. Joyce went out searching for bulls every day and found them in places where I did not usually go. She discovered definite bull areas, where single bulls or small groups of bulls spent their time feeding and resting. It was here that Joyce came across the bulls that "disappeared." It seemed that for much of the year the large bulls spent time away from females in all-bull groups. It was beginning to look as if we had something similar to the sexual cycle that Iain had suggested. What was particularly intriguing was that GP seemed to have some relation to the cycle.

In the next six months I continued to collect data on GP and estrous behavior among the females. I saw several more bulls with GP and each time they were with females and often they were with a female in estrus. Joyce came out for Christmas vacation and spent two weeks in Amboseli. By now we were calling the phenomenon "green penis syndrome" instead of "disease." She was very excited about the data that I had collected during her time away and felt sure that something significant was going on. Just before she returned to the United States, she had lunch with Harvey, who by coincidence had just received a paper on Asian elephants. He handed it to her to look at. Joyce leafed through it quickly and then suddenly stopped dead, staring at one picture, and said, "Oh my God, that's it!" The paper was on the phenomenon of musth in the Asian elephant and the picture was of a bull in full musth. Arrows pointed to his streaming temporal glands and to his wet legs and dribbling urine.

The paper described the characteristics of musth, a Hindi word referring to the physiological and psychological state that male elephants come into periodically. The most obvious characteristics are the swelling of, and copious discharge from, the temporal glands, the continuous discharge of urine, and a sharp rise in aggressive behavior. Musth bulls have always been notorious among domestic working Asian elephants and in circuses and zoos throughout the world. They have to be isolated and chained up and are usually given reduced feed until they go out of musth. Further studies in Asia showed that musth is accompanied by high levels of the male hormone testosterone in the blood.

In the Asian male elephant musth is confined to those bulls that have reached puberty. An individual male often comes into musth around the same time each year and it usually lasts for two to three months. The occurrence is unsynchronized; that is, males do not all come into it at the same time. In wild Asian male elephants there is a positive association with female herds during musth. Scientists who studied it in the Asian elephant concluded that the phenomenon is similar to rutting behavior in deer, with the main difference being the lack of synchrony. There is a definite mating season among deer, during which all the adult males go into rut and the females come into estrus. Female elephants, however, come into estrus throughout the year.

When Joyce read the paper Harvey gave her and some other related papers, everything began to fall into place. She came back in the summer of 1978 and followed both musth and non-musth males. Bad Bull, right on schedule, came into musth in June, July, and August for the third year in a row, and gave Joyce a very bad time. She was extremely frightened at first and came back to camp white-faced and shaken on many occasions, but Joyce is one of the most determined people I know and she continued to go out with the elephants. If Bad Bull saw her with a herd of cows and calves he would come for her car, sometimes from over 400 yards away. Often she would leave but circle around and watch from another point. If he was in consort with an estrous female he was usually unable to chase Joyce because if he left the female the other males would race in to claim her. He would continue to glower at Joyce and threaten her if she came close enough, but under these conditions Joyce could make observations from a distance. Fortunately, some other bulls, who were more tolerant of her presence, were also in musth that summer and she was able to collect valuable data on their behavior.

By the end of the summer of 1978 Joyce and I knew that musth occurred in the African elephant, but to understand the phenomenon more fully a detailed study needed to be carried out. Joyce went back to college, wrote up the work she had done on the bulls thus far for her honors thesis, and graduated in June 1979. Like me, she was completely hooked on elephants and was totally fascinated by their complex lives. She registered for a Ph.D. at Cambridge University to work under the well-known animal behaviorist Robert Hinde, whom I had met in Manyara and kept in touch with over the years. With grants from the New York Zoological Society and Smith College Joyce came out to Amboseli in January 1980 and carried out an 18-month field study on musth.

At the end of 1980, with five years of accumulated data on individual bulls, we wrote a joint paper for the journal *Nature* announcing the discovery of musth in the African elephant. The big question our colleagues and others wanted to know was why it had never been noticed in African elephants before. There seemed to be several reasons.

The main confusion centered on the temporal-gland secretions. In the Asian elephant, primarily males secrete and only when they are in musth. Females very rarely secrete. African elephants, on the other hand, secrete from the temporal glands frequently and the females more often than the males. Females secrete in all seasons and throughout the year, but most frequently in the dry season. We and others have speculated that the females use the secretion as a means of communication. It may be a way of keeping members of families and bond groups in touch when they are fragmented while feeding in thick bush or at considerable distance from one another. The secretion probably has an individual odor, and elephants, with their excellent sense of smell, might use it to find one another. When a single elephant is separated from its family it almost invariably has fresh temporal-gland secretion. Calves of both sexes from about six months old also secrete, but as males get older they do so less often, and as full adults they usually secrete only when they are in musth, and then the liquid seems thicker and of a different consistency. Since temporal-gland secretion is one of the main signs of the onset of musth in the Asian elephant, when observers in Africa saw females and calves with secretions, they concluded that it served a different purpose and that musth did not occur in the African elephant.

Other elephant observers, including Iain, had seen males dribbling urine, but they had not associated it with a specific physiological state. Most of the other people who had studied African elephants had examined the carcasses of elephants shot on culling and cropping schemes. Very few scientists had watched individually known live animals over several years. Joyce and I believed that this was the key to why we were able to discover musth. We watched known bulls, with GP and temporal-gland secretions, appear in the areas used by cows, exhibiting sexual interest in females and acting aggressively toward males, and then a few months later we saw the same bulls with no external symptoms quietly spending their time with males. When we observed this for several years in a row we could not help but see a pattern. As it turns out, once we described the phenomenon in the African elephant, people started seeing it in elephant populations in other places.

The next question after the simple discovery of musth was what its function and significance was. No other mammals exhibit this type of phenomenon. In most species in which females are sexually receptive at all times of year, males pursue or at least guard females throughout the year (e.g., lions, zebras, baboons). In those species in which females are receptive for only a short breeding season males in turn are sexually active only during that time (e.g., wildebeests, deer). It is a generally accepted theory that an individual's main goal is to pass on its genes to as many offspring as possible in the course of its lifetime. Animals use different strategies to try to achieve that result, but a first look at the sexual-activity cycle of male elephants would indicate that they were acting counterproductively to the goal. Females are in estrus all the year in Amboseli, with the majority of them in estrus from January through July. But a big male like Bad Bull, who theoretically could be pursuing females from January through July, was actively pursuing them only during June, July, and August. Why? And what possible advantage could dribbling and smelling have for a male?

Joyce set out in her Ph.D. study to try to come up with some of the answers. It would take a whole other book — and someday I hope Joyce will write it — to talk about male strategies, but I want to describe some of the things that Joyce found and her theories as to why they happen. Joyce has basically concluded that musth evolved as a result of competition for females. In polygynous species, in which males compete intensively to mate with females, there is usually a marked sexual dimorphism, with males growing significantly bigger than females. Male elephants (at 11 to 12 feet at the shoulder and 12,000 pounds) are almost twice the size of females (at eight feet at the shoulder and 6000 pounds), which suggests that in the evolution of this species bigger males were more successful in mating with females and they then passed on their genes to their male offspring, who in turn inherited their fathers' large size, and the biggest among these sons again produced more offspring. Eventually, given intrinsic and ecological constraints (such as a limit to the amount of food a huge animal could find and consume), male elephants got as big as they could. Musth may then have evolved to add another dimension in the competition for females.

In Amboseli we divided adult males into six age classes: 1a) 11–15; 1b) 16–20; 2) 21–25; 3) 26–35; 4) 36–50; and 5) 50+. Males reach sexual maturity at about twelve years old, at which age they are capable of producing viable sperm, and theoretically could mate and impregnate a female. In the wild it would be a very rare event. Neither the adult females nor the older males would tolerate them. Young males usually leave the family in which they are born when they are in Class 1a, but they tend to stay near cow/calf groups and tag along with a few other young males at the periphery. Joyce found that gradually, as males get older, they spend more time in all-male groups in "bull areas" away from the cows and calves. Bulls do not usually begin to compete for estrous females until they enter Class 3, and even then they can only hope to sneak matings when the bigger males are not around or are distracted. The Class 4 and 5 males are more successful at mating. Bulls begin to come into musth at around 30 years old, and the older males in musth are the most successful of all.

Among the bulls Joyce detected dominance patterns based on size and therefore age. Each male appears to know his relative rank in relation to every other male in the population. In Amboseli the five oldest bulls are: 1) M13; 2) M22; 3) M126; 4) M45; and 5) M99. If these bulls meet each other at, say, a fallen tree, M13 will almost always be able to supplant M22 and M126 at the best branch and so on down the line. But if one of these bulls comes into musth all that is changed. If M22 is in musth and meets M13 not in musth, M13 will back down and avoid a confrontation with him. Musth thus confers greater dominance status on an individual.

For much of the year the Class 4 and 5 males spend their time in bull areas. Certain males stay in certain areas and they always go back to these places when they are not in musth. It is a bit like an all-male club. The big males wander around with one or two buddies or on their own feeding, resting, mud wallowing, and dusting. It is a very peaceful period, during which they are gaining weight and building up their reserves of energy. When a bull comes into musth he leaves the bull area and searches for females. It is almost a Jekyll and Hyde sort of change. Everything about the bull seems to alter. First the bull's temporal glands start to swell, then to secrete, and finally he begins to dribble urine. At the same time his posture changes. He carries his head high, tucks his chin in, and moves with a purposeful, long stride. Joyce can detect a musth male from over half a mile away across a plain just by his posture. Along with the "musth walk" are some other postures and behavior. Bulls in musth wave their ears in a peculiar way, unlike the earflapping of females or non-musth males. Joyce theorizes that the ear waves waft the scent of the bull's temporal secretions before him. Musth bulls also have a special vocalization never used by non-musth males or females. It has a low frequency and probably covers a long distance. Even when the bull is nearby the sound seems muted and distant to the human ear. Bulls in musth also do odd things with their trunks, frequently raising them up to their foreheads in a characteristic posture.

However, the most striking change from non-musth to musth is in the bull's conduct. He becomes extremely aggressive to other bulls and, as we have seen, sometimes to human observers as well. Joyce had read that testosterone levels rose markedly in the blood of Asian musth bulls and she was keen to see if the same thing happened in African bulls. It was impractical to try to get frequent blood samples from wild elephants, but it was possible to collect urine from the Amboseli elephants. Harvey and I had done it in the early part of the study. Most of the Amboseli soil is fairly hard in the dry season and when an elephant urinates the liquid does not immediately sink in but leaves a puddle for a brief time. Harvey and I were collecting urine from females and it was easy enough to drive up to them and gently ask them to move away so we could get out and obtain a small amount. It was another thing to try to get urine from a musth bull, but if anyone was going to do it it was Joyce. In her tiny Suzuki jeep she would race up to a bull after he urinated and somehow convince him to move off a few yards; then she would jump out and quickly collect the urine in a huge syringe.

The urine was transferred to bottles, labeled, and stored in a freezer. It was then shipped to Bill Lasley and his team at the San Diego Zoo, where it was analyzed. The findings were fascinating. Bulls in musth had markedly raised testosterone levels. Testosterone has been linked to aggression in other mammals including man, so the behavior of the musth males was not surprising. Nor was it surprising to find that musth bulls will seriously fight one another.

Fighting was known to be rare among elephants. Iain had seen only one fight in his time in Manyara. I did not see a true fight until 1978. I had witnessed young bulls and even fairly large males sparring and sometimes getting quite heated in the process, but a fight is a different thing altogether. When two bulls spar it is basically a friendly testing of each other's strength and these tests form and maintain the relative dominance positions. In sparring the two males usually move right up to each other and touch trunks to each other's mouths and sometimes to the temporal glands, then they may entwine their trunks, or put them on each other's heads. After this gentle investigation they push their heads together, interlocking their tusks. In this position they might shove each other back and forth or back up and rush together. Sparring does not escalate into a fight.

In true fights the two bulls do not touch with their trunks first. They approach each other with heads high and ears out and there is a lot of maneuvering and sidestepping to keep lined up. They then make swift, short, sharp lunges at each other, coming together with immense force. Sometimes they lock tusks and each tries to twist his opponent off his feet. If one bull falls the other will tusk it through the head or side and kill it. A fight is a frightening thing, and I still get very upset watching one. Joyce has seen scores of fights, and although she can remain calm enough to take notes and film them with a video camera, she always finds them distressing, especially when she thinks that one of the bulls may be killed. On several occasions she has heard the horrible sound of a tusk cracking and seen as much as three quarters of a tusk drop to the ground. Some of the fights have gone on for at least eight hours. Circumstantial evidence leads us to believe that two bulls in Amboseli have been killed as a result of fighting in the last few years.

Watching a bull in musth over the two or three months he is in that state gives a good indication of why males are not in musth year-round. The musth strategy seems to be for a bull to put all his energy into pursuing females for a few months out of the year. During musth the bull spends most of his time traveling in his search for females. When he finds a female, he can spend very little time feeding or resting because he must chase away other males and guard the female. On some occasions he may have to fight another musth bull. Joyce took a series of photographs to assess the condition of musth bulls and found that the bulls gradually lost weight over the musth period until at the end they were very thin, and at that point they went out of musth and returned to their bull area.

Thus, for example, Bad Bull cannot afford to be in musth all year round. Nor can he come into musth at just any time of the year. Much depends on when bulls that are dominant to him come in. Bad Bull comes into musth in June just when M13 is losing condition and going out of musth. June, July, and August are not the best months to find estrous females; March, April, and May are better, but then Bad Bull would have to compete with M13 and M22, who are both dominant to him. He would be risking his life and would probably mate with only a few females that the other two had not found. Later in the year he is the dominant musth male and has the field to himself.

In the end, musth appears to give a male great advantages in his relations with other males. Non-musth bulls are capable of mating but they will always give way to a musth bull in any dispute. Thus, if a musth male finds an estrous female he will usually be able to mate with her and then guard her during the crucial time when she is ovulating. Joyce speculates that the dribbling of urine, the secretion from the temporal glands, and the characteristic postures, gestures, and vocalizations of the musth bull constitute a visual, vocal, and olfactory advertising. A bull in musth is announcing that he is in excellent condition, is surging with testosterone, and is willing to fight, and fight to the death, for estrous females or simply for his continued status. (Elephants are long-lived and what happens in future years may be more important than a single estrous female.) Other elephants can see, hear, and smell a musth male before he even arrives where they are. Depending on an elephant's sex, the animal may respond in a variety of ways to the knowledge that there is a musth bull in the area. For example, when a musth bull wanders about he leaves a scent trail from his dribbling. Joyce has found that males and females react differently to the scent trail. Females will stop and sniff a trail and will usually show excitement by rumbling and flapping their ears. The way a male reacts to the trail depends on his own musth state, who the bull was who made the trail, and their current dominance relationship. A younger male in musth, on finding the scent trail of an older, dominant musth male, will stop, smell the trail, and look around anxiously for the bull that made it. At the same time he might decrease his own dribbling and temporal-gland secretion. Usually a dominant male will simply ignore the scent trail of a younger musth male. When the males are close in rank, one may track the other in order to have a confrontation and settle their dominance relationship.

Both my data and Joyce's on mating behavior indicate that females prefer mating with musth bulls and that they may actually exercise some choice in the matter. In a paper I wrote for the journal *Behaviour* entitled "Oestrous behaviour and female choice in the African elephant," I described estrous behavior and set out my reasons for believing that females had choice and why they chose as they did.

From my data I showed that a female, if determined, can on most occasions outrun a male that is pursuing her. Thus she does not have to mate with every male that chases her. When a female goes into consort with a male, allowing him to mate and guard her, in 87 percent of the cases it is with a musth bull. In the first section of this chapter I described how Tia was able to outrun her pursuers. She eventually mated with a large male in musth, Bad Bull, and went into consort with him. After a couple of days he lost interest in her, presumably after testing her urine and determining that the period in which she was ovulating had passed. By the time he left she had probably already conceived. This was a typical series of events.

In the *Behaviour* paper I speculated that there were several advantages to a female choosing to mate with a large Class 4 or 5 male in musth. There are two immediate short-term advantages. First, by choosing to mate and consort with a musth male a female may be able to avoid harassment by other bulls, since the musth male will be able to chase off and threaten other bulls who approach her. By staying close to a musth male a female reduces the stressful consequences of being in estrus. Being chased by males for three or four days appears to be exhausting, disrupts family cohesion, reduces the normal time spent feeding and resting, and could cause injuries to the female or her youngest calf. Second, with only a few days in which to conceive, it is important for a female to find a fertile male. It has been shown in other mammal species that high testosterone levels can correspond to increased production of sperm. Musth bulls, with their high testosterone levels, may, therefore, be more likely to impregnate a female.

There may also be long-term advantages to mating with a musth male. Only the older males come into musth, that is, those over 30, and the ones that are most successful and stay in musth the longest are those over 40. An older male has in a sense proven that he has "good genes" simply by surviving. In addition, if longevity is heritable, it would be advantageous for a female to pass this trait on to her sons and daughters. Longevity is important for female elephants, since the longer females live the more calves they can produce and raise successfully, while longevity may be even more important for the reproductive success of males, because they do not start to father calves regularly until they are well over 30.

It might be argued that a female does not really have much choice in whom she mates with because she almost always has to mate with the bull who happens to be in musth at the time she is in estrus. At times there may be more than one bull in musth around an estrous female, but it would be unlikely that she could choose between them. The outcome of the interactions between the males would override the potential choice of the female. However, I think it is worth considering that a female might come into estrus in response to a particular male being in musth and in that case she may be exercising choice.

Personally, I cannot imagine why Tia wanted to mate with Bad Bull, but then she may have seen something in him that I did not.

The T families did not change in number or composition during 1978, but given a 22-month gestation period I expected quite a few new additions in 1980.

Chapter V

Social Relationships

1979

Teresia and her two calves, Theodora and Tolstoy, and her grandchildren, Tallulah and Tim, walked south, first through a strip of *Acacia tortilis* trees and on through an area of lava rocks and small bushes. It was early morning, an unusual time for them to be moving in that direction. In the late afternoon of the previous day they had left the Longinye swamp with Slit Ear and her family, had crossed into the Ol Tukai Orok woodlands, and fed on the palms and *Acacia xanthophloea* trees for a while before settling down to sleep in an open area near the southern tip of the Enkongo Narok swamp.

Slit Ear had woken earlier than Teresia and had turned back toward Ol Tukai Orok and Longinye in the traditional pattern of moving out from the swamps at night and back to them in the daytime. She and her family were nearly a mile away when Teresia woke and stood up. The rest of her little group got up as well, stretched and yawned, and dusted and scratched themselves. Teresia gave a long, low rumble, the contact call that basically says, "Here I am, where are you?" She held still,

121

listening, her head high and ears extended, but heard no answer. She called again, this time making an even lower-frequency sound that traveled farther. She heard a distant answer from Slit Ear, and turning her head to locate where Slit Ear was, Teresia rumbled again and once more got an answer.

However, instead of heading off in the direction Slit Ear had taken, Teresia turned toward the south. The rest of her family remained facing north. Teresia swung her right front foot back and forth in the characteristic elephant gesture of indecision. She took a few steps forward, picked up some dust, and threw it on her back, called again, waited and listened. Then she decisively lifted and spread her ears, flapped them against her neck and shoulders, let them slide down with a rasping sound, and strode off toward Kilimanjaro. Her family loyally followed.

The TD family walked south for about a mile and a half without stopping to feed. Teresia seemed to have a particular destination in mind. Just before dawn they arrived at a wide depression more or less at the base of the mountain. This area consisted of small, thorny trees, *Balanites glabra* and *Acacia nubica*, interspersed with some grasses and low bushes. It was a favored feeding area for the Amboseli elephants in July and especially August. These were mid dry-season months when the more palatable and nutritious grasses in the bushed grasslands were finished but there was still good forage left in some of the woodlands outside the park. The elephants used the woodland to the south before finally retreating once again to the swamps within the park for the late dry-season months of September and October. While feeding there they had to walk back to either Longinye or Enkongo Narok swamps to drink, but they usually did this at night. Teresia, who was in her late fifties, had been using this area in July and August from the time she was a young calf. Following the tradition of her family she had decided to spend a few days in the south.

For three days the TD family fed in the depression, drinking at night or in the early morning, but they were never completely relaxed there. There were many Maasai around, and the elephants had to be alert to their movements and activities. Still the food was good and nothing unpleasant happened. But the loss of Trista had reduced the size of the TD family, and they did not feel as secure as they would have in a larger group with just that many more eyes and ears and trunks on the alert. Without the eight members of Slit Ear's family for company and added protection, they were isolated and vulnerable. On the morning of the fourth day Teresia led her family north, toward Amboseli and Slit Ear.

When they crossed the park boundary and found themselves back in the Ol Tukai Orok woodlands, the younger members of the family started running and trumpeting. Tallulah, Theodora, Tim, and Tolstoy let their bodies go loose as they ran, which made their ears and trunks flop about. At the same time, they trumpeted with a loud pulsating sound associated with play. Teresia more or less maintained her dignity but she too broke into a run. When the younger animals got ahead, Tim and Tolstoy stopped, whirled around, and came together for a sparring match. Tallulah and Theodora found some bushes and began bashing through them and beating them with their tusks. Their tails were curled up over their backs, their heads were held high with their noses poked out, and the whites of their eyes showed as they rolled the pupils up. Teresia caught up with the others and stood with her head up and out. In her usually placid brown eye was a decidedly wild glint.

The young animals played for another ten minutes before everyone began to feed on the regenerating acacias, breaking off small branches and carefully stripping them of the tasty and nutritious bark. After a while they switched to the grass that was growing under and around the trees. It had been a year of unusually high rainfall and even though it was getting toward late dry season there was still some grass remaining. They fed for another 20 minutes and then all stopped at once when they heard an elephant give a contact call about a mile away. They listened for a moment but resumed feeding and did not answer because it was not someone from their bond group.

Soon after, they began walking slowly, feeding as they went, through the mixed acacia and palm trees, finally arriving at the northern edge of the woodland at midmorning. Here they rested, standing in a tight bunch under one of the yellow-barked trees. On reaching this spot Teresia called several times but got no response. She and her group were eager to rejoin Slit Ear, but they did not know where she was.

After their rest they crossed the pan and open plain and

reached the northern end of the Longinye swamp. They fed there for the rest of the day. Teresia continued to call at intervals, but still got no response. All the members of the group had fresh temporal-gland secretions almost continuously. They met several other family groups in Longinye but ignored them, passing them by or feeding near them without interacting.

That night they moved out of Longinye to the east into the Olodo Are woodlands. They spent the night feeding and resting. Teresia called several times during the night and in the early hours of the morning when the group woke up. On that occasion she heard an answer more than a mile away toward the northwest but she was not sure it was Slit Ear. She started to move in that direction in any case. They walked west until they arrived at an area of regenerating acacia trees on the eastern side of Longinye. Here Teresia stopped and rumbled. She instantly got a response from about a quarter of a mile away and to the south. This was definitely Slit Ear, and Tia and Tess answered as well.

Teresia changed her course and began walking rapidly to the south. She and the others were streaming from their temporal glands and the sides of their faces showed thick black lines all the way to their chins. They walked through the patch of young acacias and emerged in an area of swamp interspersed with wooded islands. At a little rise above the swamp, they stopped and Teresia, Theodora, and Tallulah rumbled, listened, heard a response, and rumbled a louder sound. They were clearly excited, with heads held high and ears lifted. They shifted their course slightly and, almost running, plunged down into the swamp course and up the other side onto the first island.

Suddenly ahead of them was a group of elephants running out of the trees and coming straight for them. Teresia stopped for a moment in alarm, then recognized Slit Ear, and both groups ran, rumbling, screaming, and trumpeting toward each other. The younger animals had moved ahead of Teresia, but when the two groups came together, Slit Ear ignored the others and pushed through them to reach Teresia and greet her. Both elephants raised their heads up into the air and clicked their tusks together, wound their trunks around each other's while rumbling loudly, and holding and flapping their ears in the greeting posture. They whirled around and leaned and rubbed on one another. Meanwhile all the other members were greeting each other with much spinning, backing, urinating, earflapping, entwining of trunks, and clicking of tusks. All the elephants were producing so much temporal-gland secretion that it dribbled down along their chins and into their mouths. Above all, the sounds of their greeting rent the air as over and over again they gave forth rich rumbles and piercing trumpets of joy.

After 18 years of watching elephants I still feel a tremendous thrill at witnessing a greeting ceremony. Somehow it epitomizes what makes elephants so special and interesting. I have no doubt even in my most scientifically rigorous moments that the elephants are experiencing joy when they find each other again. It may not be similar to human joy or even comparable, but it is elephantine joy and it plays a very important part in their whole social system.

Compared to most other animals elephants lead remarkably complex social lives. Even in primate societies, which are generally considered to be highly intricate, an individual's relationships rarely extend beyond the group it is born into or at most to neighboring groups. For example, in the baboon social system, the females stay in the group or troop in which they are born, and within that troop they have very complex relationships. They may encounter individuals from neighboring troops but their interactions are brief. Males leave their natal troop and transfer to other troops in the area and thereby increase their social relationships, but over the course of a lifetime a male baboon would still interact with a restricted and relatively small number of individuals. Elephants also live in female-bonded groups in which females stay in their natal group for their lifetime and males leave at maturity. However, both male and female elephants' relationships radiate well beyond the family group through a multi-tiered network of relationships encompassing the whole population.

The network of relationships is the fiber that makes up elephant society. These relationships range from the strongest bond of all, that of mother and infant, to mere acquaintance-ship between animals whose ranges rarely overlap. I find it easiest to conceive of the system by thinking of each individual as being at the center of a series of concentric circles, each circle getting larger as it gets farther from the individual (see diagram). If, in 1979, I take Slit Ear as the central individual,

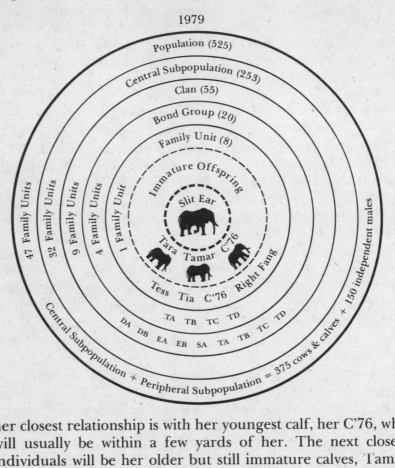

her closest relationship is with her youngest calf, her C'76, who will usually be within a few yards of her. The next closest individuals will be her older but still immature calves, Tamar and Tara. I consider these four animals to be in the first circle. In the second circle are the other members of her family, Tia and Tess (who is probably Slit Ear's older daughter), and Tia's offspring, her C'76 and Right Fang. Slit Ear will rarely be more than 25 yards from these animals. Sometimes there may be a split within the family but it will usually last only for a few days. Data collected over the years on association patterns between the individuals within the family reveal that Slit Ear and her calves are found together 99 percent of the time and Slit Ear and the other family members are together over 90 percent of the time.

More conspicuous than the percent of time these animals spend together, which is an accumulative observation, is the behavior among the individuals. The TC family usually acts in a coordinated manner, with everyone feeding, drinking, traveling, or resting at the same time and taking their cue from Slit Ear, the matriarch. The close bonds among the individuals are constantly being reinforced by tactile interactions and vocal communication. Members of a family are often in physical contact, rubbing on one another, touching each other with trunks, leaning against one another when resting, and for the younger animals, frequently playing. Older female calves spend a considerable amount of time taking care of the younger calves, and in this way bonds between different-aged animals develop early on. Even when they are not in a tight-knit bunch, Slit Ear's family maintains contact with gentle rumbling vocalizations, calling back and forth to each other when they are spread out in thick vegetation.

But the best indicator of the strength of the bonds between elephants is the greeting ceremony. Almost all elephants will greet one another but the nature and intensity of the greeting depends on who the elephants are, what their relationship is, and if it is a close one, how long they have been separated. If Slit Ear's family joins Estella's while feeding in Longinye some of the family members might greet with an exchange of trunks — that is, put the tips of their trunks in each other's mouths. Usually a subordinate animal will be the one to initiate the greeting. It almost always occurs among the younger members of the two families and there are no accompanying vocalizations. Slit Ear would not deign to greet anyone in this manner; nor would Estella for that matter, so they would basically ignore one another while the other members of their families quietly greeted each other.

A greeting within the family, however, is a very different thing. Once in a while Tia will break off from Slit Ear and feed on her own with her two calves. It may be an accidental separation or it may be deliberate on Tia's part. Slit Ear can be very competitive over food and Tia always gives way to her. But for whatever reason they separate, which is not often, their reunions always involve a special ceremony. If they have been separated for only a short while, say a few hours, then the greeting will be at low intensity. Tia might join Slit Ear by walking up to her and lifting her head, raising and flapping her ears, and rumbling the throaty "greeting rumble." Depending on her mood, Slit Ear will usually answer in the same way, overlapping with Tia's rumble. They might then reach trunks toward each other. If Tia has been separated for a few days then the greeting ceremony will inevitably be far more intense and carried out with great energy and excitement. The two subgroups of the family will run together, rumbling, trumpeting, and screaming, raise their heads, click their tusks together, entwine their trunks, flap their ears, spin around and back into each other, urinate and defecate, and generally show great excitement. A greeting such as this will sometimes last for as long as ten minutes. I believe that the greeting ceremony maintains and reinforces the bonds among the family members.

Family members are by definition related to some degree, but I would be tempted to give my right arm to go back in time in order to find out exactly how they are related. With some families I felt fairly confident about working out their family trees, especially if it was a relatively small family. For example, in 1979 I drew family trees for the TA, TB, and TD families that were probably close to being accurate. The TB and TD families were very simple. Tania was the mother of three male calves each separated by about four years. Teresia was probably the mother or grandmother to everyone in her family. The TA family had only one question. Torn Ear was the mother of her C'74, Tilly, Teddy, and Tuskless. Tonie was the odd one out. It appeared that Tonie's mother died early in the study, but I wondered how that female was related to Torn Ear. Could she have been her daughter or sister or what?

The TC family was a little more difficult. Slit Ear could have been Tia's mother; she was estimated to have been born in 1936, while Tia was estimated to have been born in 1950. That would have made Slit Ear 14 years old when Tia was born. Females can give birth at 14 (the youngest female to give birth in the Amboseli population was 11) so theoretically Slit Ear could have been Tia's mother, but I intuitively doubted it. I suspected that Tia may have been the daughter of an older

With Torn Ear was a young tuskless female about 11 years old.   HARVEY CROZE

The upcurved female with the slit in her right ear was called Slit Ear.

At the rear of the group was a straight-tusked, ancient but immensely dignified female who was named Teresia.

A very "handsome" female who became known as Tania had the youngest calf in the group, a male only a few months old, who several years later was named Toby. HARVEY CROZE

A large female with V-shaped notch out of her right ear became known as Torn Ear. HARVEY CROZE

Teresia had a daughter about 11 years old who was
given the name Tina.

The young female with
a big V notch out of
each ear became
known as Tallulah.

A "pretty" young adult female who was eventually called Tia had a funny little calf with her
who had one left tusk and a tiny stunted right
tusk, and he was immediately named Right Fang.
HARVEY CROZE

Members of the T bond group meet in Longinye and perform an intense greeting ceremony. HARVEY CROZE

A large male in musth, M13—Iain, comes striding over with head high and ear waving to supplant the younger male.

M13 stops and tests the female's urine by touching and smelling it with his trunk, and then placing his trunk in his mouth to assay the hormone levels.

Finding the results of interest, he sets off in pursuit of her.

Joyce following the musth male Iain, who is the biggest bull in the population

Gentle greeting/sparring

A young female from one family greets a small calf from another family.

Two male calves playing:
The favorite game of
calves is to climb on a
sleeping animal.

The one on the
bottom wiggles to
get the other off.

He struggles to
stand.

He gets up and turns to
meet the challenge of the
other.

Estella and her twin calves, Eclipse and Equinox, who were born in June 1980

An older sister gets a sleeping calf up onto its feet when it is time for the family to move on.

Wes inspects the damage after an elephant raid on the kitchen.

The author,
Masaku, and Moshi
NICOLAI CANETTI

An elephant near death who was speared and became ill with septicemia

An elephant who was speared by the Maasai; she carried the spear in her head for over a year before it finally fell out and she is now fine.

A young female speared by Maasai warriors proving their bravery; rangers have opened up the carcass so that lions can more easily feed on it.

Vultures descend on a dead elephant.

Agatha of the AAs stops and feels and gently moves her mother
Annabelle's skull, even though it is several years after Annabelle's
death and Agatha passes the spot frequently.

Elephants will inevitably inspect the carcass or skeleton of a dead elephant they have not come across before: Here the members of a family touch and stroke and move the bones of an elephant that died in the park; the female on the right tosses a bone in the air.

Teresia at 58

Teresia at 51 years old and
Tolstoy at 2 years old
HARVEY CROZE

Tolstoy at 15

Slit Ear at 37 and
Tara at 6

Slit Ear at 50
(with a recently broken tusk)
and her C'83

Tara at 19

Tess at 10

Tess at 23

Right Fang at 3

Right Fang at 17

Tuskless at 13

Tuskless at 24

Teddy at 7   HARVEY CROZE

Teddy at 21

Author and first research vehicle in 1975   MARION KAPLAN

Author and Land-Rover in 1987   JOYCE POOLE

female in the family who died, possibly the previous matriarch. Tess, who was born in 1963, was probably Slit Ear's daughter, but I had no positive evidence except that she stayed close to Slit Ear and the gap between her age and Slit Ear's next youngest calf was four years, the average calving interval. Tara, who was born in 1967, was following Slit Ear and they were acting in the typical mother-calf way when I first met them in 1973, and so it was highly likely that she was Slit Ear's daughter. Because Tamar, born in 1971, was still suckling from Slit Ear in 1973, there was little doubt that she was Slit Ear's calf. With Slit Ear's youngest calf, her C'76, there was no doubt. Once the study began, the maturity of any calf that was born was positively known. Tia's offspring presented no problem. Right Fang was still suckling in 1973 and the birth of Tia's C'76 was recorded on the very day he was born.

To complicate my efforts to sort out kin relationships within the family, there was the question of paternity. With an average birth interval of four to five years, and given both the attrition rate of large adult males and the musth phenomenon, in which an individual male is only sexually active for a few months of each year, the chances of subsequent calves of one female having the same father are probably not high. Thus Slit Ear's daughters were probably half sisters. On the other hand, because of musth, the chances of calves born within a few months of each other in one family having the same father (but different mothers) are fairly good. In the previous chapter I mentioned that Slit Ear was in estrus only a few days before Tia and that she too had been in consort with Bad Bull. My data show that there tends to be a synchrony of estrus within families — that is, the females who are available (those who are neither pregnant nor suckling calves under two years old) often come into estrus within a few weeks of one another. If there is one dominant bull in musth at the time when a family's females are coming into estrus then that bull will probably father all the calves conceived in that period. The calves will then be half brothers or half sisters on their father's side, and half brothers or half sisters to their siblings on their mother's side, and grandchildren, nieces, nephews, aunts, uncles, or cousins to some of the same and other members of the family. In fact, the relationships are probably more than a little convoluted.

Thus the animals in the first two inner circles of the diagram will be closely related under any circumstances, which is an important consideration in understanding behavior between individuals. In the third circle from the center are the members of Slit Ear's bond group. Closest to her are the TDs, Teresia's family, and then the TAs and TBs, led by Tuskless and Tania. All four of these families form close bonds, exhibited by the amount of time they spend together and their behavior both on meeting and when they are together. The association data show that Slit Ear's family, the TCs, are found with the TDs over 80 percent of the time. That is a very high level, indeed, one of the highest of any of the bond groups in the population. Evidence of the strength of the bonds between these two families is substantiated by the intensity of the greeting ceremonies they perform when they come together after a separation. In the first part of this chapter I described a greeting between the TCs and TDs. These two families are particularly close and it appears the main tie that binds them is the very strong attachment between Slit Ear and Teresia. By their ages, they could be mother and daughter but I have always suspected it is another relationship. Whatever it is I am sure it goes back the full 40-plus years of Slit Ear's life.

Slit Ear is not as closely bonded to the TAs and TBs. From a few greetings I saw in the early days of the study I believe that Slit Ear had a fairly close relationship with Torn Ear. After Torn Ear died in 1976, however, the attraction between the families appeared to weaken and they began to spend less time together. The TCs spend 34 percent of their time with the TAs and 35 percent of their time with the TBs. This is still much higher than their association with any other family in the population. The next closest family is the DB family, at only 15 percent. All the other families in the population are found with the TCs at even lower levels of association.

The relationships among the families and between the individuals within a bond group intrigue me. I believe the animals are related but at what level I do not know. My birth records go back only to 1970 so it is impossible for me to know how, or even if, the adult females are related to one another. With regard to the present adult females, it is a question that can never be answered, but I should get some insights in the years to come as I watch what happens to the structure of the families and bond groups in which I do know the relationships.

One event has already helped my understanding of how bond groups are formed. In the first years of the study the biggest family unit in the population was the PAs, led by the matriarch Penelope. There were several other large females in the family as well: Phoebe, Patricia, Philomena, Priscilla, and Zoya (definitely a P member, but originally given a "Z" name because she was first found in a small subgroup that I mistakenly thought was a new family unit). The PA family grew to 29 members, which was very large, considering that the average family unit was only nine. The family broke into subgroups from time to time but there did not seem to be a definite pattern to the separations until 1978–79. More and more often the family split into two groups, one section led by Penelope and the other by Phoebe. Previously the family had not seemed to be able to decide who was going to line up behind Penelope and who behind Phoebe. In the following two years they appeared to work it out, with 16 animals forming Penelope's group and 13 forming Phoebe's group. By 1982 I considered them two separate family units, the PAs and the PCs. Nevertheless, they still spent more time with each other than with any other families in the population; when they met they performed the greeting ceremony, and when they were together they moved in a coordinated way and showed affiliative behavior toward each other's members. In other words, these two families formed a bond group.

I suspect that the division of the PA family is typical of the way bond groups originate. Family units simply get too large to be efficient social groupings, and either they split more or less down the middle or a subgroup breaks off. In Lake Manyara Iain watched the gradual breaking off of a small subgroup of five animals from a large family of 22. The splinter group consisted of an adult female, her young adult daughter, and their three calves. The leader of the subgroup was relatively young compared to some of the other females in the family. Iain suggested that competition from these older females in a large family led the younger female to leave so she could forage on her own with only her closest relatives. But as in Amboseli, her newly formed family maintained close bonds with the original family and often rejoined them or moved in the same area as they did.

The fourth circle out from Slit Ear includes the members of the families that make up her clan. Now I have to admit that I am not sure the concept of "clan" means anything at all to Slit Ear. As I explained in Chapter III, during the dry season particular families tend to go back to particular areas of the Amboseli basin. The four T families spend the dry season concentrating on the Longinye swamp and the Ol Tukai Orok woodlands in the daytime, and the lava scrub to the south and the Olodo Are area to the east at night. They are very rarely seen using the Enkongo Narok swamp and the area west of that swamp. There are five other family units that more or less follow the same pattern: the DAs, DBs, EAs, EBs, and SAs. I consider that these nine families make up one clan. They share the same range amicably and sometimes even move together in aggregations, but they reveal no close bonds with one another. I have never seen a T animal approach any of the individuals from the other families and perform the greeting ceremony.

I have, however, seen aggressive interactions between clan members. One animal will approach an individual or group from another family with ears folded horizontally in a characteristic posture of aggression, and the individual or group will either move away, stand its ground, or respond aggressively in turn. It can happen when one animal wants something the other is feeding on, or wants the shade of a tree the others are standing under, or it may happen for no reason apparent to a human observer. Aggression also occurs within families and bond groups. In fact, aggressive interactions occur most frequently within families, which is not surprising because the members are together most of the time. There is a drop in aggression within bond groups and clans, and an increase again between individuals from different clans. Slit Ear is often aggressive to other elephants, and some of my colleagues say she is a "real bitch." She is, but I like her spirit.

In the fifth circle out from Slit Ear is her subpopulation. In Chapter III, I described the division of the population into central and peripheral subpopulations, with the central elephants concentrating on the basin swamps in the dry season and the peripheral elephants ranging in the bushland surrounding the basin and using the swamps only for drinking and for some periods of feeding. Slit Ear and the other Ts are members of the central subpopulation.

Slit Ear's subpopulation consists of 32 families; 26 of these families overlap with her in and around the Longinye swamp in the dry season and she will meet them from time to time. The other five families spend more time in the region of the Enkongo Narok swamp and so she will probably not meet them in the dry season more than a few times. Her relationships and interactions with the other members of the subpopulation depend on whether they are family members, bond-group members, clan members, overlapping clans, or more distant clans. She will know all the other elephants in the subpopulation and she will know her relative dominance position to them. On the whole her interactions with her family and bond group will be amicable, although she will compete with the other individuals and might even give someone a poke in the backside for not getting out of her way fast enough. Her interactions with the rest of the subpopulation will be few and far between in the dry season, but she may interact aggressively with those animals she does meet.

Slit Ear's final circle of relationships consists of the whole population, including the peripheral subpopulation and the adult independent bulls. Many of the animals in these two categories she will not meet at all during the dry season. The peripheral subpopulation, consisting of 15 families, spends the dry season west of Enkongo Narok and south toward the mountain. Slit Ear's home range does not overlap with theirs except possibly in some of the night feeding areas to the south. Generally, it is only when the rains begin and the system of dry-season home ranges breaks down that the two subpopulations meet. Then the family units join together in large aggregations and move throughout the whole basin and surrounding areas.

When the family units from the different clans and subpopulations meet and merge, there always appears to me to be a heightened social awareness with a great increase in interactions. Male calves in particular seek out the young males from families they do not know well and spar and test each other's strength. The adult females, too, test their dominance relations with other females, threatening or sometimes even chasing another female, to reestablish old, or determine new, relationships. Slit Ear is one of the highest-ranking females in the whole population and she does not miss many opportunities to make sure everyone acknowledges this. At the same time there are many amicable and playful interactions between members of different families. Elephants seem to be attracted to other elephants, and when the ecological conditions allow it they appear to prefer to be in these large, socially stimulating groups.

Clans and subpopulations may only be a figment of my imagination in that I may have superimposed a structure on a simple geographical separation of families. On the other hand, the animals may perceive themselves as belonging to a particular dry-season home range, which they are willing to defend from intrusion by other families. It will take years of data collection to find out whether there is exclusion of some families from other families' home ranges, but I strongly suspect there is. My colleagues and I have seen incidents where a group from one area has come into another group's home range and has experienced aggression from the residents. If one area is better than another, and there is evidence to suggest this, then it might be reasonable to speculate that the residents would try to defend that resource. This is one of the questions that makes me want to keep the study going for many more years. From a theoretical point of view, in the field of behavioral ecology, the question of the defense of resources by kin and non-kin and the forming of coalitions in that defense is an important one, but it is equally important for the management of elephants in conservation areas. It is essential to know what might happen to an elephant population if part of its range is cut out and some of the population has to immigrate into new areas, or if some of its members are culled and certain areas are left without residents.

Finally, but not inconsequentially, are Slit Ear's relationships with the 150 adult males in the population. Her attitude toward them depends on several factors: the age of the male, his kin relationship to her, her own sexual status, and — if he is an older male — whether or not he is in musth. In the dry season Slit Ear will generally encounter males on a daily basis. Young males, ranging in age from 10 to 20, may join her family for feeding or traveling and she may or may not tolerate their presence. My impression is that Slit Ear is fairly intolerant of young bulls. Other families seem more welcoming and a few of the families always seem to have a little gaggle of males following along behind. Except when they irritate her in some way, young males are almost totally ignored by Slit Ear and it could be said that she has almost no relationship with them.

Older males of approximately 20 to 30 years are more problematical. By their early twenties most males will have become larger than Slit Ear and therefore dominant to her. By mere bluff and past experience she may still be able to see off a 20- to 22-year-old but eventually she will give in and accept his new status. However, these transitional males do not go around fighting with females — it is not in their interest. More commonly there will be a showdown in which the male and female will stand with their heads high and their ears spread and try to make the other back down. In these confrontations with females, the males are probably feeling their way into their new status as socially mature bulls. In general, though, males 20 to 30 years old have already started spending more of their time in the company of other males in the bull areas. During the dry season they will be separated from the females for much of the time, and Slit Ear will meet these bulls only when their feeding ranges overlap or if the bull comes into the cow-calf areas searching for estrous females. If Slit Ear was in estrus she would try to avoid a male of this age while waiting for an older male to come along.

Bulls over 30 have entered a whole new realm, that of the musth male, and there is no question whatsoever of their status. Slit Ear would never try to challenge any of these older bulls. More likely she will greet a large male when he comes into the group. If he is in musth she will probably show some interest and excitement and she and her family might respond to his arrival with a chorus of rumbles. Again I would give an arm or a leg to know what Slit Ear's relations are with the large males in the population. She is older than most of them (due to poaching and hunting of males for their far bigger tusks, there are many fewer males over 40 than females; only two or three males are over 50 while there are several females well into their fifties and a few may even be in their sixties), but some could be younger brothers, nephews, or cousins. Does she treat these individuals any differently from non-relatives? Would she mate with a brother, and what may be more interesting, since she does not have much choice with a musth male, would he mate with her?

In the wet season Slit Ear will contact and interact with males more frequently than in the dry season. Males of all ages are found in the aggregations that form when abundant vegetation is available. Typically adult males will make up about 10 percent of an aggregation. Thus if there is a group of 300 elephants, there will be around 30 adult males. If there is a female in estrus the proportions are likely to be different. I once found a group of 200 elephants that included 63 adult males. There were two females in estrus, one of whom was young and inexperienced, and it was utter chaos.

For long stretches of time Slit Ear basically will be uninterested in males. When she comes into estrus it is another matter. Then she tries to attract them and keep them interested in her. Being an older, very experienced female she will not tolerate younger males mating with her and will avoid them until she finds a musth male to go into consort with. Since I have known Slit Ear, she has had several calves three to five years apart in age. (An average calving interval of four years means that a female conceives, gives birth after a 22-month gestation period, followed by about a two-year period of not ovulating, and then is ready to conceive again.) Slit Ear was probably in estrus only one time for each of the conceptions I recorded, which means that, on average, once every four years, for a few days, she found males of considerable interest. Aside from her own estrous periods, Slit Ear will show some interest in males when her daughters and other close family members come into estrus. Then she will join in the demonstrations and commotion that occur when a female is pursued and mated. But again this will not be a common occurrence in the course of a year. In general, a female's interactions with the adult males in a population will be perfunctory and of major importance only once every four years or so.

Thus, in summary, Slit Ear's relationships radiate out through the six social circles, from intimate strong bonds with close kin to mere acquaintanceship with peripheral elephants and most of the males most of the time.

During 1979 I had many opportunities to see the social relationships at work. It was a year of high rainfall and lush, abundant vegetation and the elephants were often in large aggregations where many interactions occurred. I saw the four T families between 30 and 40 times each during 1979. They were not around as often as some families but my sightings of them were frequent enough for me to keep up with what they were doing. The realignment of bonds seemed to have solidified. Slit Ear and Teresia's families were almost always together, and Tania and Tuskless's were nearly inseparable. Tuskless had definitely taken over the role of matriarch for her small family. I was very pleased to see that during this time of abundance Tania and Tuskless had at least temporarily abandoned the garbage pits and become "real" elephants again, moving away from the central part of the park and traveling with the large groups.

It was generally a good time for all the elephants, although the Ts suffered a loss early in 1979 or late in 1978. Sometime toward the end of 1978 Tania's older son, Taabu, disappeared. At first I thought he might have gone independent. He was nine years old at the time, which would have been young for striking out on his own but not unheard of. But by March 1979, when I had not seen him with either his family or other families or bulls or on his own, I had to conclude that he was dead. There was very little poaching or spearing at the time, but young males around his age are vulnerable. Taabu was a particularly cheeky young male, frequently getting into mischief around the lodges and campsites (he once picked up a small tent of Joyce's and carried it away, waving it about in his trunk, with me shouting and chasing him until he dropped it), and perhaps he got too sure of himself around some Maasai who did not appreciate his antics. On the other hand he may have eaten something in the garbage pit that killed him, but I never found his carcass.

I was away in England and the United States for six weeks in May and June 1979. I was always very reluctant to leave the elephants, knowing I would be missing important events, but this time I had another colleague on the project. Keith Lindsay, a young Canadian ecologist, had come to work on the feeding behavior and habitat use of the elephants in 1978, and it was reassuring to know that he was there in Amboseli keeping track of the elephants. Although Keith did not have to know all the individuals for his ecological work, by the time I left he knew most of the adult females and many of the independent males.

Keith's study was an integral part of the overall research project on the Amboseli elephants. From the beginning of the study, Jonah, Harvey, and I had wanted to find someone to work on the ecological aspects of elephant biology, in particular on how the elephants used the Amboseli habitats and how they interacted with those habitats. Next to cattle, elephants were the single most important herbivore in the ecosystem and potentially they could cause significant changes in the vegetation by their feeding activities. We needed someone to look at these activities.

It was not until 1978 that we found someone suitable. Joyce was back in the States for most of 1978 and all of 1979 finishing up her undergraduate degree and starting her Ph.D. program. I was sharing my camp with Phyllis Lee, a Cambridge Ph.D. student who was working on the social development of infant vervet monkeys. Phyllis was a smart and lively Californian whose company I enjoyed tremendously. This arrangement worked out well, but I was still keen to have a second person on the elephant project. Thus I was very pleased to have Keith join me.

Keith had done some research on bears in Canada and then grew restless and came to Africa on an overland trip. When he reached Kenya he looked up various ecologists he had heard about, one of them being Harvey. Harvey put Keith in touch with me and I in turn put Keith in touch with Jonah and we all came to an arrangement whereby Keith worked for Jonah part-time on the ecological monitoring of the park and for the rest of the time on his own study of the elephants. Keith registered for a master's degree at the University of British Columbia and was sponsored through Jonah by the New York Zoological Society and through his own grant from the Natural Sciences and Engineering Research Council of Canada. He stayed in Jonah's field house, about three miles from my camp.

As with Joyce's, Keith's arrival marked the beginning of a long and fruitful collaboration on the Amboseli elephants. From the beginning Keith was an enthusiastic, hardworking, and very dedicated researcher. In addition, and somewhat unusually, Keith looked well beyond the science to the greater context in which the elephants had to live — to conservation in general and the problems involved in trying to integrate the needs of people with the needs of wildlife. This concern of Keith's and his ability to articulate it well made him an extremely valuable member of the project.

When I got back from my trip in June I was fascinated to hear from Keith how everyone was. Studying elephants is a bit like watching a soap opera. I wanted to know what was happening, who had had babies, who had been in estrus or musth and, always with a holding of my breath, who had died. Keith had mostly good news for me. Many new babies had been born, some to females I had seen in estrus, which confirmed that the behavior I had been recording was reproductive behavior. Keith also had some sad news for me. Alyce had been speared in May, apparently when the AAs were out of the park. Her young calf, who had been born in April, died with her, and Amy's new calf, born in May, disappeared at the same time. In the week before I returned Tuskless's C'76 disappeared. This was difficult to understand, for the calf was a healthy three-year-old and it was a very good year in terms of food. I suspected that her death was garbage-related. By 1979 the TAs and TBs, except for brief periods at the height of the wet season, were very serious users of the garbage pits and they also raided the campers persistently and thoroughly. Perhaps Tuskless's calf ate something poisonous or a sharp piece of glass or large plastic bag (the latter two items have been found in elephant dung near the lodge). I was disturbed by this because I hated seeing these two families so addicted to human foods. The situation was becoming acute, particularly in the campsite where two or three camps were being raided day and night. I thought the park authorities might feel that they had to shoot some of the Ts.

Nineteen seventy-nine seemed to be a year of deaths indirectly related to human activities. The week I got back to Amboseli I began to smell something dead near the camp. The smell got stronger and stronger and it had a particular characteristic that made me think it was a dead elephant. I drove out in my Land-Rover to the public campsite area nearby and searched through all the glades among the palms and acacias. Finally I broke through into a hidden glade and found a gruesome sight. The headless decomposing carcass of a young female elephant sat in an abandoned garbage pit. Somehow the head had been severed and carried away about 20 feet, almost certainly by lions or hyenas. From the evidence it looked as if the elephant had fallen in the pit and then could not get out. She may have broken a limb or injured her spine in the fall. There were indications that other elephants had tried to help her. They must have given up eventually, whereupon the lions or hyenas came in and I imagine attacked her while she was still alive, killing her and feeding on her head, which was above the lip of the pit. The tusks were still intact so I knew it had not been poachers. I notified the park rangers and they came with a tractor, covered over the carcass, filled in the pit, and collected the tusks.

The ears of the elephant had been eaten so I could not tell immediately who it was. Finally, through censusing the families and seeing who was missing and by comparing the tusks to those in photographs, I discovered that it was Pamela of the PA family, Philomena's 17-year-old daughter. Poor Pamela. It was an unnecessary and cruel death and I felt angry that it had been caused by human negligence. But, I told myself, there were bound to be some adverse effects of tourism and its related human activities in the park, and, everything considered, Amboseli's wildlife was little affected by the presence of lodges, camps, vehicles, and roads. That year just seemed to bring a spate of incidents: the growing problems with Tuskless and Tania's families, the disappearances of Taabu and Tuskless's C'76, and the horrible death of Pamela.

Otherwise I came back to a very good period in Amboseli. It was the third year of exceptional rains and the park was looking green and vibrant with vegetation. In the camp the grass was over three feet high and yellow-flowered bushes and purple-flowered shrubs had grown up that I had never seen before. The elephants were fat and healthy and many new babies were being born after almost two years of no births. It was such a joy to go out each day in anticipation of what I would find: Would it be a glorious aggregation of over 500 elephants moving across the plain, or a young female whom I had known since she was a young calf coming into estrus for the first time, or a tiny new calf barely able to walk?

Back in 1978, I had already begun to check the breast development of all the females who were old enough to conceive. Elephants have two breasts located between and just behind their front legs. In shape and position they are very similar to human breasts. Males and little females just have small nipples on a flat surface. Lactating females have large, swollen breasts with prominent nipples. By noting the size and fullness of a female's breasts each time I saw her and comparing them to a lactating female's, I discovered that females gradually develop enlarged breasts as their pregnancy progresses. This turned out to be useful in determining pregnancy in some of the females. It was a good indicator for those females who had never had calves before or for those females who had lost calves or whose older calves were no longer suckling. Females such as Slit Ear and Tia, who were still suckling their '76 calves, had huge, swollen breasts so it was impossible to use the size and fullness of their breasts as pregnancy indicators. I thought they were probably pregnant because I had seen them both in estrus in 1978, but there were no physical signs. Usually females over 20 years old do not appear to bulge or get fatter when they are pregnant. Younger females sometimes look noticeably fatter and rounder toward the end of the pregnancy.

On February 26, 1979, I checked the breast sizes of the females in the TC and TD families. I noted that Slit Ear's and Tia's were "full" and that Tallulah's breasts had begun to enlarge and were already bigger than Tess's. I suspected that Tess had been pregnant once before and miscarried. She had small, slightly developed breasts, about one eighth of the size of a lactating female's breasts, but with no fullness. On the other hand Tallulah's breasts were about one-quarter size and starting to fill out.

In July 1979 I noted further changes in some of the T females. Tallulah's breasts were now one-half size with definite fullness. Tania suddenly seemed to have large, full breasts. On July 1, I noted: "Tania has three-quarter sized, full breasts — very distinct change."

In August Tuskless and her gang were being particularly bad about raiding the campers and on August 12 the rangers had to come over and shoot above their heads. I did not see Tania again for about ten days and thought she had been frightened by the shooting, but then on August 23 she appeared in the camp with a large, healthy newborn male calf. He was the center of a great deal of interest in the family. Tilly and Tonie fussed over him, reaching their trunks to him and touching and fondling him. He appeared to be enjoying the attention greatly.

Phyllis, my camp-mate, was enchanted with this baby, whom she named Tom, and fascinated by all the behavior that revolved around him. It was about then that I sowed a seed in her mind. I wanted her to come back to Amboseli to study calf development in elephants when she had finished her thesis. I also plotted a little in another direction by encouraging a burgeoning romance between Phyllis and Keith. I wanted Keith to come back after he finished his master's thesis, and thought how nice it would be for them to do their studies together.

The rest of 1979 was fairly uneventful for the Ts. They continued to enjoy very good conditions. There was no poaching and few incidents with Maasai. As I look back now, I wish every year could be as good as this one was for them. One lovely late afternoon with the sun slanting through the palms, I found the Ts near my camp. In fact I drove out because I heard the sounds of elephants playing, in particular the nasal, pulsating play trumpet that they make with their heads down while running, charging, or beating through bushes. I found Theodora, Tara, and Tamar playing among the palms with several young males, including one of my favorites, Conrad, who had recently left the CB family. Their play was very active, with the young females chasing and sparring with the young males, all with a great deal of noise. I took some notes but mainly just sat there enjoying them. I wrote:

1730 Theo and Tara still being very silly, chasing males. Theo comes running out of the palms almost straight into me. She demonstrates, lifting her head, spreading her ears, making brief dancing lunges at me. Then she finds a log and attacks that, first with her front then with her hind feet, kicking it and breaking it up by stepping on it. She charges me flat out, trumpets, circles around, backs up, runs off. I start the engine to reposition the car and she comes tearing back and gives an impressive charge with lots of noise. She's wonderful.

Theo, Tara, and Tamar played for well over an hour, expending a tremendous amount of energy. Tallulah, who was usually one of the most playful and sprightly of the young females, did not participate. On December 12 I noted, aside from her full breasts, that she actually looked pregnant. She was fat and there was a definite bulge. She also seemed to be moving more slowly. I felt quite excited about her pregnancy. She was one of the females I had known since she was an adolescent and it was oddly fulfilling and rewarding watching her become an adult female.

At the end of 1979 there were 20 animals in the four T families:

TA	Tuskless (62)	♀
	Teddy (66)	♂
	Tonie (67)	♀
	Tilly (70)	♀
TB	Tania (44)	♀
	C'79 — Tom	♂
	Toby (73)	♂
TC	Slit Ear (36)	♀
	C'76	♀
	Tamar (71)	♀
	Tara (67)	♀
	Tess (63)	♀
	Tia (50)	♀
	C'76	♂
	Right Fang (70)	♂
TD	Teresia (22)	♀
	Tolstoy (71)	♂
	Theodora (67)	♀
	Tallulah (63)	♀
	Tim (69)	♂

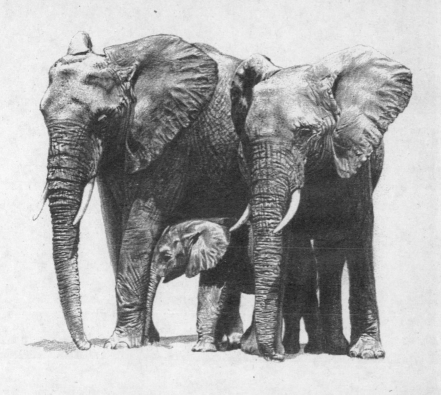

Chapter VI

Births and Babies

1980

Tallulah lay down to sleep with the other members of her family in the early hours of the morning of March 30, 1980. She had difficulty lying down and once she was down she could not rest comfortably. After a short time she got up with great effort, first rolling onto her stomach and then rising on her knees and finally onto her feet. She stood with her head hung low, her trunk tip resting on the ground. She dozed but did not sleep as soundly as the others.

Half an hour later Tallulah abruptly lifted her head, flapped her ears, and rumbled. Teresia stirred but did not get up. Tallulah kicked her foot in the dirt and after loosening the soil she dusted herself. She walked about ten yards from the group and fed halfheartedly on the grass at the edge of the open area where the group was sleeping. After about five minutes she resumed dusting. Then she walked back to the others, who were still lying in great heaps on the ground, and tried resting in a standing position. This nap lasted only about 20 minutes before Tallulah swung her foot, scratched her eyes and her

147

ears, and picked up the fine soil and threw it over her back. Her family continued to sleep, some of them so heavily that they snored.

Tallulah was restless for the remainder of the sleeping period, which lasted until around 5:00 A.M. Just as the sky was beginning to lighten, her family, together with Slit Ear's, stirred and one by one got onto their feet. They stretched, yawned, scratched their eyes and ears with the curve of the trunk just above the tip, and, almost as if on signal, all urinated and defecated with loud sounds of whooshing and plopping.

Tallulah was already fully awake and feeding about 15 yards away. She came over and greeted them with a gentle rumble. Her behavior was not in synchrony with theirs, and some of them reached their trunks out to smell her. She turned and backed into the group in a gesture of submission and nonaggression. Theodora rubbed her head against Tallulah's backside and Teresia placed her trunk on Tallulah's vulva and gave one sharp shake of her head. Tallulah rumbled again and Theodora and Teresia answered.

After some more dusting and scratching and greetings and friendly rubbings on one another, the group moved north toward the southern end of the Enkongo Narok swamp. They stopped and fed for varying lengths of time, gradually working their way to the swamp. Tallulah plucked up a few bunches of grass from time to time, but did not feed in earnest. Every once in a while she suddenly held still for several seconds. At one point she shook her head and walked backward a few steps. She then stopped, swinging one leg back and forth.

By 8:00 the two families had passed along the southeastern edge of Enkongo Narok, heading toward the Ol Tukai Orok woodland. They were crossing through an area of tall *Sporobolus consimilis* grass and briefly halted to feed on the new shoots of this tough, coarse grass. By now Tallulah was not eating at all and was definitely behaving unusually. She kept hesitating and backing up and generally acting agitated and disturbed. The other elephants appeared to be ignoring her.

At 8:20 Slit Ear made the "let's go" rumble and set off across the open pan toward the woodland. All the others followed. As they neared the edge of the woodland Tallulah suddenly dropped down on her back knees. Just under her tail a slight bulge appeared. Her vulva hung down lower than usual and liquid was dribbling from it. She got up and walked into a thick clump of mixed acacia and palm trees. There she lay down on her side for about five minutes. When she stood up, the bulge that had first appeared below her tail was much bigger and had moved down about a foot. In the next few minutes Tallulah was very restless, moving in and out of the clump of trees and backing and turning around. The bulge slipped down farther.

Suddenly, at 8:44, with no apparent extra effort on her part, Tallulah gave birth. The baby elephant, completely enveloped in a fetal sac, was expelled and on the ground in a matter of seconds. Tallulah remained quiet for about 45 seconds and then turned around to the sac and gently touched it with her foot. Inside the calf kicked its legs and then was quiet. The other members of her family were about 20 yards away feeding. Tallulah scraped the grass and earth around the sac with her front foot in an apparent effort to clear the ground. She reached down and tried to free the baby from the sac by using her tusks. The calf began kicking again and Tallulah scraped the ground more frantically.

At 8:47, three minutes after the birth, some of the younger T females — Tara, Tess, Theodora, and Tamar — came over with heads up, rumbling and threatening the small vehicle which was nearby. The females crowded around the mother and infant, vocalizing loudly, and joined in helping to remove the sac. Just then a young independent male, M170, pushed his way in to Tallulah. This upset Tallulah and she threatened everyone, chasing the young females away. The females in turn attempted to chase away M170, but he would not be intimidated. A second, larger male, M12, entered the clump. By now Tallulah and the other females were very agitated and had streaming temporal glands.

At 8:54 Slit Ear arrived, threatening everyone. Some of the others rumbled and urinated at her approach. She turned and backed into the clump. Tallulah was trying to lift the baby, which was now free of the sac, with the aid of her foot and trunk. Finally, at 8:56, 12 minutes after it was born, she got it to its feet, but it was unable to stay up. Two minutes later it stood again, but fell down. In the meantime the males, stimulated by the unusual smells and behavior, kept poking and shoving Tallulah and laying their trunks and tusks along her back as if they were about to mount. She ignored them as much as possible, holding her ground and not letting them push her away from the calf. Tallulah, however, seemed to be confused about what to do with the baby, and kept digging at the ground, pulling up and trampling vegetation, breaking branches and backing and turning.

Around 9:10 several more females joined the group around the newborn calf. These females were from the VA family, which was not bonded to the Ts. Some of the larger ones threatened the young T females, causing a great commotion. There was much aggressive posturing back and forth between the families. In their excitement two young females got down on their back knees and raised their heads in the air. All of them crowded around Tallulah and the baby, and one of them found the fetal sac, picked it up, put it in her mouth, tossed it around, and finally threw it over her head.

After ten minutes of intense activity the VAs and most of the T females had gone, leaving Tallulah and the baby with Tara. At 9:21, 37 minutes after birth, the calf was finally more or less able to stand. It searched for its mother's teats but ended up trying to suckle from Tara and fell down again. It got up and fell down several more times, but finally at 9:43, 59 minutes after the birth, the calf was able to walk. Tallulah, the very wobbly baby, and Tara came out of the clump of trees. The baby had difficulty negotiating a root but Tallulah helped it over. Just then Right Fang arrived and took it upon himself to defend Tallulah and the calf from the small vehicle that was still there. Out in the open the calf appeared to gain strength and although it had to stand with its legs spread wide in order to keep its balance, it was now able to take a few steps without falling down. It reached its trunk out, smelling and searching, flapped its ears, and gave a shake of its head. Tallulah reached down and touched the umbilical cord, which still hung down, then gently felt the baby's head, face, and mouth.

On that same morning in March 1980, I was sitting at my desk in my tent analyzing data when a safari vehicle arrived. I reluctantly left my work and went out to see what the driver wanted. He told me that my colleague Joyce Poole had sent him to tell me that "an elephant had been born in Q-18" and I was to come immediately. I thanked him profusely because it was very kind of him to go out of his way to come to the camp. I leapt into my Land-Rover and started out to the other side of Ol Tukai Orok, wondering where in the one-kilometer grid square, which was designated by the coordinates Q-18, Joyce might be. It took me less than ten minutes to find her.

Joyce had gone out to look for bulls that morning and had spotted the TCs and TDs at 8:25 crossing the open plain to the south of the woodland. She noticed that Tallulah was behaving oddly and so she stopped and watched the group for a while. She was very fortunate to see the birth and all the behavior around Tallulah and the new baby from very close up. The elephants were only about twenty yards away and sometimes closer.

I arrived shortly after 10:00 and took over the observations from Joyce. Although several hundred calves have been born in Amboseli during the course of the study and there have been times when three observers in three different vehicles have been out with the elephants, only two births have ever been witnessed by researchers working on the project and very few anywhere in Africa. Possibly the majority of births occur at night and perhaps those that do take place in the daytime happen in secluded places.

By coincidence I had seen the other birth five days before. I will describe that one before I continue with the story of Tallulah and her calf because it presents an interesting contrast. On March 25 I was watching a large aggregation of elephants down at the far wet end of the Longinye swamp where I could not take the vehicle. I was parked about 300 yards from the elephants. They were all peacefully feeding in the long grass when, at about 12:00, my attention was attracted to some earflapping and rumbling vocalizations from members of the DA and DB family units. Dinah, an adult female, was streaming with secretions from her temporal glands and Deborah, another adult female, was standing in an unusual posture: Her head was up, tail out, legs spread, and from under her tail, halfway down between her hind legs, was a large bulge. She appeared to be straining. After about 30 seconds she stepped back with legs still spread and reached her trunk into the grass.

Dinah, Daisy (a 12-year-old female), Denise (Dinah's five-year-old daughter), and Daniel (Deborah's three-year-old son) all came running over and also put their trunks down into the grass. I could not see much from where I was, but Dinah and Deborah seemed to be working at something with their trunks and feet. After 12 minutes I saw a wet, shiny object wiggling in the grass. Two minutes later it struggled up and there was a baby elephant.

Deborah, an experienced mother who was approximately 47 years old and had had several calves, appeared relaxed and competent from the moment the baby was born. The calf fell down five times in the first half hour, but Deborah carefully got it up by gently prying a foot under it and steadying it with her trunk. The calf, a female, continuously explored its new environment, reaching its trunk out, touching and smelling its mother and other group members. The young females, Daisy and Denise, were particularly fascinated with the new calf, feeling it with their trunks and standing over it in a protective manner. Daniel came over and felt his new sister once and then apparently took no more interest.

When the calf was only 34 minutes old Deborah resumed feeding; after 40 minutes the calf took its first shaky steps forward; and after one hour and 36 minutes it found Deborah's teats and suckled vigorously for over two minutes. Later it even suckled from Dinah and was completely tolerated. Both Deborah and Dinah stood quietly with the leg near the calf well forward so it could reach the nipple. When I left three hours after the birth the calf was walking well and steadily behind Deborah.

The contrast between Deborah's behavior with her new calf and Tallulah's was striking. Tallulah was about 17 years old at the time and had not had a calf before, but I was still surprised when Joyce described Tallulah's very disturbed state. While Deborah had appeared unperturbed by the birth, Tallulah was distinctly upset and apparently confused about what to do.

By the time I arrived, Tallulah, Tara, and the baby had gone deep inside another clump of young acacia trees. It is common for an adolescent female to accompany a new mother, but I would have expected it to be Theodora, from Tallulah's own family, rather than Tara, Slit Ear's 13-year-old daughter, who stayed with Tallulah.

When Tallulah had been out in the open earlier, Joyce had had a good look at the baby and determined that it was a female. After relating the events up to that point, Joyce left to try to find the bull she had set out to observe earlier and I started recording behavior at least every five minutes using the time of birth, 8:44, as the theoretical starting point:

1029 I can hear the baby making suckling sounds. They are deep within trees and bushes and I can't see anything.

1034 Tallulah is backing and kicking dirt. Can see a little now. Violently kicking ground.

1039 Continued breaking up of vegetation and trampling by Tallulah. Tara remains relatively quiet.

1044 (two hours after birth) Still trampling. Clearing area? Heard suckling sounds again but must be on Tara because Tallulah too active.

1049 Calf is reaching and exploring Tara. Tallulah is still trampling.

1052 Calf vocalizes — sort of a hoarse growl.

This type of behavior continued for another half hour. I could see that the calf kept reaching out its trunk and exploring the bodies of the two females. Several times it made sucking motions and sounds against the side or leg of one of the females, but it had not yet found the teats.

At 11:19, when the calf was just over two and a half hours old, they came out into the open again, only a few yards from my car. I could see the calf clearly. A baby elephant is almost a perfect miniature of an adult except that it has no tusks and it tends to have more hair. This calf had black hair on its forehead, very red, bloodshot eyes, and the backs of its ears were bright pink. Its hind legs were very stiff and it had difficulty walking, but like Deborah's calf it continuously explored its new environment with its trunk. Tallulah and Tara were alert, disturbed, and restless, and both continued to kick the ground, loosening earth and throwing dust over themselves.

1120 Calf walks to Tar. Tal turns and follows. Tal backs toward calf. Tar turns, sways, head up. Both alert. Trunks up, very nervous. Low rumble 2×. Trunks up again.

1124 Tal backs around to get beside calf which is standing with head to Tara's tail. They are standing in the full sun. Calf tries to suckle from Tar. She moves forward. Again Tal backs around so calf beside her. It reaches toward Tar. Umbilical cord still hanging.

1129 Tal turns around, calf still facing Tar. Tar reaches toward it with trunk. Tal and Tara sway. Tal turns around and reaches trunk to baby's face 2×. Calf's legs are wobbly as stands. Still in sun. Tal turns and maneuvers near it. Reaches for umbilical cord then shakes head gently. Calf appears to be resting. Has been still.

1134 Tal pulls calf toward her. Calf standing, wobbling on front legs, trunk on ground. Certainly seems weaker than Deborah's. Tal moves away from calf. Tar turns and reaches trunk toward baby. It is definitely resting, in Tara's shade. Now Tar moves forward and Tal backs toward calf. Tal reaches trunk to umbilicus 2×. Possibly tried to pull bit off.

1139 Baby still resting. Tar moving around. Tal feeding.

1144 (3 hours after birth) Bit of blood on the ground. Tal picks it up and eats it. Tal pushes Tar out of the way 2×. Baby still searching, exploring. Very wobbly. Tal is hopeless, moves every time and does not put leg forward.

The calf was now over three hours old and I was sure that it had not yet suckled. The two females seemed totally inept. At this point I do not think the calf knew which female was its mother and it was wasting a lot of time trying to find Tara's teats. Tallulah tried to maneuver so that the baby was beside her but she did not have the experience to direct the calf to her teats and then stand quietly with her leg forward so it could suckle. She may have been delivering the afterbirth as blood still dripped from her vulva, and this may have been one reason why she was so restless. In any case she was rarely still for more than a few seconds. Each time the calf got near Tallulah's teat the kicking motion prevented it from finding the nipple. One time it almost got its mouth on the teat but promptly got hit in the nose and was nearly knocked over. I found it very upsetting and frustrating to watch.

In the next half hour the baby continued to try to find Tara's or Tallulah's teats. At one point it did find Tara's teat and suckled for about a minute but Tara had no milk. They remained out in the hot sun and March is the hottest month of the year in Amboseli. At 12:17 the calf gave up searching and began to rest out in the full sun.

1221 Baby lies down. Tal starts kicking the grass all around it. It struggles, back legs in air. Tar also kicks dirt. Tal goes around, feels with trunk, kicks it while kicking grass. It sits up, then lies down again. It's still in the sun. I'm worried that it is going to get dehydrated.

1224 Both are kicking grass all around it.

1227 Tal goes to the other side of calf and gently kicks it. It struggles but doesn't get up. Now she pushes on it.

1229 Still kicking dirt around and on calf and on themselves.

1234 Calf struggles but doesn't get up. In sun again.

1239 Calf almost completely buried in dirt and grass. Both Tal and Tar making kicking motions around it. It's breathing oddly. Struggles then lies still.

By now I was almost as frantic as Tallulah and Tara. The calf seemed weak and unable to get up by itself and Tallulah did not know how to pry the calf up with her foot, at the same time steadying it with her trunk, in the way that Deborah did. When they had nearly buried it I began to give up hope and assumed I was watching the death of a newborn calf caused by the inexperience of its mother. I wanted desperately to do "something" but knew I could not and would not interfere. My palms were sweating, my stomach was in a tight knot, and I wanted to shout at Tallulah to tell her what to do, but, of course, it would not have done any good.

Then, at just about 12:44, when the calf was four hours old,

Tallulah went around behind its head and put the tip of her foot under its shoulder, and the baby, making a hoarse rumbling noise, struggled to its feet. It walked over to Tara and started searching for her teats again. It then moved to Tallulah and searched all around her body and finally, at 12:51, when it was four hours and seven minutes old, it found its mother's teat and suckled for one minute and five seconds. I felt like cheering and clapping.

I watched them for another hour until they were just beginning to walk slowly in the direction that the other Ts had taken. In that hour Tallulah continued to move about, backing and sometimes kneeling and straining, but she only produced small bits of blood. I think she must have been delivering the placenta in pieces. The calf explored, rested, tried several times to suckle again, but was unsuccessful, and frequently uttered the funny, deep baby rumble. It got stuck on a branch once and Tara disentangled it. It came over to my car and bumped against it three times. The females came over with it and did not seem the least alarmed. I felt privileged to have their trust. At 1:44, when it was five hours old, it looked stronger and was fairly secure on its feet. Tallulah was feeding steadily for the first time. I left them at 1:50 and drove back to camp feeling totally wrung out and exhausted and not at all sure the calf was going to make it.

Late in the afternoon of the same day, Tallulah, the calf, and Tara arrived in the open glade at the southern end of my camp. It meant that the baby had walked over a mile. Tania and Tuskless's families joined them and there was great excitement as they greeted the females and inspected the baby. The next morning I found Tallulah again and was pleased to see the baby suckling well and looking strong. I spent the rest of the day with them, doing special observations on the mother and calf.

In the afternoon Tallulah, the calf, and Tara joined the whole T bond group. I was confused for a moment, however, when I saw a second tiny newborn calf. I soon discovered that it was Tia's. My notes read:

1650 Tia with a new baby, a male. This calf is strong, suckles well. Has bloodshot eyes, not hairy. Still wobbly.

Backs of ears pink. Exploring, walking over to a male. Other group members aware of it, back toward it. Tia has no blood, so must have been born last night or yesterday.

I noted that Tia's older calf was not with her and wondered where it was. A little later in the afternoon I found both Tia's C'76 and her older son, Right Fang, with Teresia several hundred yards from the others. It has been suggested that the old matriarchs act in a sort of grandmother role and it certainly looked as if Teresia was playing that role on this day.

The next day, which was April 1, I set out early to try to find the Ts again in order to continue my observations on the two new calves. I searched all around the Ol Tukai Orok woodlands and finally drove over to the Longinye swamp, and there, after searching for more than three hours, I found them with yet another new baby. This one was Slit Ear's, and although I was expecting her to have a calf, it was unusual to have three calves born in a bond group in three days. Slit Ear's calf, a female, was quite hairy, with black hair on its head and back. It was very wobbly and still had a bit of umbilical cord. Slit Ear had dark stains on the insides of her hind legs and was streaming from her temporal glands. I estimated the calf to be only a few hours old.

The young females in the bond group — Tilly, Tonie, Tess, Tara, Tamar, and Theodora — were almost beside themselves with the excitement of three newborn calves. Although Tania's calf, Tom, had been born in August 1979, there had been no other babies in the other three families since 1976. While the interest by the females in these new calves was intense, the juvenile males showed very little curiosity at all about the babies. The young females kept trying to get as close to a baby as possible. Tia and Tallulah were fairly tolerant of their advances and attentions. Slit Ear allowed her older daughter Tara to stand over the baby and fondle it, but when Theodora tried she threatened her. The babies seemed to bask in all the attention. They freely went from female to female and tried to suckle from anyone. They would reach up and suck from the nipple of the flat-chested, undeveloped breast of an adolescent female and seem to derive satisfaction from the suckling alone. The young female would stand very still with her leg forward, reaching back from time to time to touch the baby's mouth, and if I could ever say an elephant had an expression of total bliss on its face it would be on one of these little females. After the baby finished I often saw the female reach with her trunk and gently touch her nipples as if in astonishment that they had some function. Presumably this was all good practice for being a mother one day. Perhaps for Tallulah it had been such a long time since anyone in her family had had a calf that she truly did not have enough experience.

By the time these three T babies were born, I was almost getting a bit blasé about new calves. Baby elephants seemed to be popping up all over. But there was a reason for this profusion of calves in Amboseli. The baby boom was a result of the severe drought in 1975 and 1976, during which the elephants almost ceased reproductive activity. Only five conceptions took place in those two years. However, a few months after the drought ended in early 1977, females renewed sexual cycling. Most of the adult females in the population were "available" to conceive and many did in 1977 and 1978. Twenty-two months later, in late 1978, 1979, and 1980, the babies appeared.

On November 25, 1978, I found Delia, whom I had seen in estrus in February 1977, with the first tiny, newborn calf. I may have been as excited as her whole family was. I too had not seen a new baby for well over a year. The next month four more calves were born; in 1979, 57 calves were born; and by April 1, 1980, another 28 calves had been born, making a total of 90 calves in one year and four months.

By the end of 1979 advisers and friends kept saying, "What a marvelous opportunity for studying calf development," and I kept saying, "But I don't have the time." By then I had seven years of data that needed to be analyzed and written up in the form of scientific papers. That year the New York Zoological Society had started sponsoring me and they expected me to slow down on the field work and do a lot more desk work. I agreed that that was what I should be doing and so resisted the temptation to start a new study. Then, in February 1980, George Schaller, the well-known wildlife biologist and director of NYZS's research division, came to Amboseli to visit Jonah, who is employed by NYZS, Joyce, who had just started her Ph.D. project on the bulls with support from NYZS, and me.

We all went out to look at elephants and George took one look at all those babies and said, "You must do a study of calf development and behavior." I bowed to the inevitable and in March 1980 I put aside my mountains of data and started systematic observations on the calves. I never regretted the decision for a moment. The calves were delightful to watch and they gave me all sorts of insights into adult behavior patterns and elephant social behavior in general.

I decided to make observations on six newborn calves and six calves who were nine months old and study both sets until the end of the year. The two sets combined would span the first 18 months of an elephant calf's life. For the older set I chose three males and three females who had been born in June 1979. Of these I selected two with mothers who had never had a calf before, two with mothers who had had one previous calf that had died, and two with experienced mothers with one or more surviving calves. For the newborn set I also wanted three males and three females with mothers with a similar range of experience. In this case I had to wait for appropriate calves to be born so that I could start watching them from their first days of life.

I was delighted that the three calves from the T families fit my requirements because the Ts were usually easy to find and very nice to work around. Two weeks after Slit Ear's calf was born Tuskless gave birth, which among other things meant that four out of the six adult females in that bond group gave birth to calves within three weeks of each other, suggesting a definite breeding synchrony in those families. It also meant that I had another calf to add to my set.

I decided to study all four of these calves, because, although all lived in the same bond group, each was growing up under different circumstances. Tallulah's calf, a female, had no older brothers or sisters; Tia's, a male, had two older brothers; Slit Ear's, a female, had four older sisters; and Tuskless's, a male, had no siblings, but his mother had had one calf who had died at three years of age. The fifth calf in the group, Tom, who was not in my set but whom I watched interacting with the other calves, had had no calves his own age to play with until these others were born. However, Tom had not lacked companions, for all the young females in the bond group paid a great deal of attention to him. From the day he was born he was followed and fondled and fussed over by five young females. Possibly as a result of this undivided attention, Tom was always very social, outgoing, and apparently self-assured, spending much of his time away from his mother, Tania.

To simplify taking notes on a special data sheet, I gave each of the calves a code name. I took its mother's first two initials and the last digit of the year it was born. Thus an example of a code for a 1979 calf was LI9 for Lillian's calf. For the 1980 T calves the codes were TA0 for Tallulah's, TI0 for Tia's, SL0 for Slit Ear's, and TU0 for Tuskless's. As the zero for 1980 tended to become the letter O in my mind, these codes were also pronounceable as names. Eventually all four of the T calves' codes became their final names, which were then spelled Tao, Tio, Slo, and Tuo. Most of the other calves born in 1980 and all the calves born in other years kept their codes until they were weaned and then were given names. LI9, for instance, became Lewis.

When Tuskless's calf was born in mid-April, I completed my set of six newborn calves and the study was well under way. The two other calves that made up the set were Lynne's female born on March 2 and Penelope's female born on March 22. Lynne had had a calf in 1976 who had died and Penelope was an older, experienced mother with several surviving calves.

I learned a lot about these calves in their first year and although I sometimes cursed them and their mothers when I could not find them, I always enjoyed the sampling when I was with them. I was interested in both the physical and the social development of the calves. At birth an elephant calf weighs around 260 pounds and stands about two feet nine inches at the shoulder. Although this may sound big compared to most other newborn animals, it is actually only 4 percent of an adult female's weight and 2 percent of an adult male's. An elephant has a lot of growing to do, especially a male.

A newborn elephant is considered "precocious" — that is, it is able to walk at birth and keep up with the adults in the group. However, that is just about all it can do. It is otherwise completely dependent on the care of its mother and other family members. In the first days of life a baby elephant is almost totally helpless. It is barely able to see and apparently finds its mother mostly by smell, touch, and sound. It can walk but only just — it moves along on stiff, wobbly legs and frequently gets stuck on a small root or a clump of grass from which it has to be extricated, usually by an attentive adolescent female. Its eyes are rimmed with red and the backs of its ears are bright pink where the blood shows through the delicate skin. Even the skin of the body, although rough, looks thin and vulnerable compared to that of the older juveniles and adults. It often has hair on its forehead, head, and back and this hair is either black or, surprisingly, red. (I particularly like the little redheads because they are easy to identify within the family.) The baby's trunk is its main medium of contact with the world around it. It is constantly reaching, smelling, and touching with it. At the same time the trunk is often a problem. For much of the time the calf does not seem to know what to do with it. A baby often stands vigorously swinging its trunk back and forth, tossing it up and down, and sometimes whirling it around in a circle. The highly flexible, wiggly trunk is what makes a baby elephant look remarkably like a rubber toy. It often gets in the way. I have seen a calf step on and trip over its trunk as it is walking along. But the trunk is also a source of comfort. A calf will frequently suck on it in the same way that a human infant sucks its thumb.

In the first months of life a new calf spends most of its time resting and following its mother with brief intervals for suckling and exploring. The calf is very rarely more than a few feet from its mother and usually less than a foot from her, often touching her by leaning on her leg or by resting its head against some part of her body. If the calf strays away from its mother, either in confusion or while exploring, the mother or an older female calf will follow it.

In the observations I did called "focal samples" (because the observer focuses on one animal for a given amount of time), I noted down, among other things, who the calf's nearest neighbors were at five-minute intervals. Over 90 percent of the time the newborn's nearest neighbor was its mother and the next nearest neighbor was often a young female from the family or the older sibling of the calf, either its sister or brother, but more likely a sister if the calf had one. Older brothers tended to move off from the mother and new calf.

The mother makes soft, almost humming, vocalizations to her calf, which probably helps to keep them together. If the baby gets distressed for any reason, such as falling down, getting stuck in a bush, or getting pushed by an older animal, it will let out a loud, hoarse, squeaky-door cry that brings instant results. All the young females in the group will rush over to it and try to help, sometimes causing more confusion than assistance by getting in the way of the mother. When the calf gets older, the mother often allows the young females to deal with her calf's problem while she continues to feed. Distress calls from babies are fairly frequent, and one gets the impression that the calf is not often truly in trouble.

From the first day of life the calf explores its environment. It smells and touches and tries to pull at things; it reaches out to any and every elephant that comes near it, apparently with no fear at all. Fear is something it has to learn. Newborn calves frequently come over and touch my Land-Rover, sometimes to the dismay of their mothers or baby-sitters. In the families that are not as habituated as are the Ts, the conflict in the females is visible. They perceive the baby to be in danger but they are too frightened themselves to come right up to the car to pull the calf away. They stand with heads high, swinging a foot, or shuffling back and forth. The baby usually touches the fender or side of the car one or two times, or sometimes butts the car with its head, but soon gets bored and wanders back to its caretakers, who pull it toward them and feel it, and then give a shake of their heads or a mock charge at me in disapproval.

I was surprised at how early a new calf starts to use its trunk for manipulating objects. From the first week of life a calf begins to try to pick things up in its trunk. Sticks seem to be the favorite item. A calf is comically inept as time after time it tries to wrap its rubbery, apparently out-of-control appendage around a small stick or a blade of grass. Sometimes it gives up but often I was impressed by the perseverance and concentration that a little calf puts into a task. Eventually it lifts the stick, either grasping it with the "fingers" on the trunk tip or precariously balancing it in the curve of the trunk.

After the first week or so of life, the calf has gained strength and coordination and is able to walk well, even run. Then it can start to interact more with the other members of its family and bond group and learn how to function in its complex social world. It has to learn the protocol of interacting with siblings, other group members, strangers, adult bulls, or simply with individuals that are bigger or smaller than it is.

One day I was doing a focal sample on Tallulah's calf, Tao, and discovered how Tom had overcome the problem of being a different size in playing with the smaller calves in his bond group. The females and older juveniles were peacefully feeding in a lush area. Tom had been feeding too and then lay down to rest in the grass. Tao spotted him almost immediately and quickly approached and started to climb on top of him, a favorite game of all young calves. Tom wiggled and kicked and Tao fell off and they both wiggled and kicked with trunks and legs entwined. Tom, who was a particularly robust calf, kicked a bit harder and Tao jumped up and ran back to Tallulah. Tom got up and followed, but instead of trying to engage her in play directly, he lay down flat on his side beside her. She turned, saw him, and immediately climbed on him again. Once more they wiggled and kicked and when it became too intense Tao jumped up and ran to Tallulah. Again Tom came and lay down beside her. In all he did it four times.

Earlier in the year I had seen a large adult bull, Mark, lie down in an upright position in order to spar with another bull who was considerably smaller than he was. They had sparred playfully and briefly with both standing, but the young bull, M140, turned away, and although Mark followed, M140 would not spar with him again. Then Mark sank down on his knees with his rear legs out behind him, and as soon as M140 saw him he came straight over and started sparring. M140 was now the taller of the two. After seeing Tom and Tao playing I wondered if Mark had recalled this technique for getting a smaller animal to play from his childhood.

By watching the six babies on a regular basis over the months I was able to determine when they reached important stages in their development. Perhaps the most important was when they started to eat. It turned out to be a bit earlier than I had thought. All six calves began to spend more time fiddling with vegetation at about three months old, and between three and four months each managed to get some blades of grass into its mouth and chew and swallow them. At this point the actual amount of food they were ingesting must have been small because it took a great deal of effort to break off a blade of grass, grasp it in the trunk, and get it to the mouth. It could take well over a minute, whereas an adult could easily eat nine large trunkfuls in a minute. So often I saw a calf twirl and twirl its trunk around a single blade of grass, until finally it broke off, and then invariably it would drop it and have to spend even more time trying to pick it up. Eventually it would get the blade firmly in its trunk, lift it up, and then, apparently forgetting what the whole purpose was, deposit the grass on the top of its head.

From four months on the calves gradually spent more and more time feeding and by six months old they were getting a significant amount of vegetation in their diet. The other set of calves, which I started observing when they were nine months old, were already spending more than 40 percent of their time feeding. Nevertheless, the rate at which they suckled had decreased only slightly. They needed both the milk and the vegetation to fulfill their nutritional needs. When they reached the stage of feeding in earnest, it was interesting to watch them learning what to eat and how to handle it. A calf will frequently reach up into the mouth of its mother or older brother or sister and pull out a bit of what they are eating. I assumed that in this way the calf was learning what species of vegetation to eat. At other times the calf would stand almost under the chin of its mother or other relative and pick up the bits of vegetation that she had loosened or dropped in her own feeding. Some of the tough, short grasses needed to be kicked in order to break a clump off. The calves did not seem able to manage this maneuver and in the areas where this was the only vegetation available they had to take the leftovers from the other family members. Another technique used by the young calves was simply to kneel down and bite the grass off with their teeth. I always considered this cheating.

Around the same time that they began to feed on vegetation the calves learned how to drink with their trunks. Again this skill came much earlier than I had thought. I had estimated that it took about a year to master the technique of drawing water up into their trunks and then tipping their heads back and releasing the water into their mouths. The small calves knelt down in the water and drank with their mouths, but by three to four months they were already trying to use their trunks. Granted they spilled most of the water in the process, but at least they tried. By four to five months old they were fairly proficient at it.

It was always a delight to spend time with the calves. Needless to say I got particularly attached to the twelve that I was watching. The older ones were perhaps not quite so endearing as the little ones, but by nine months old they had developed distinct personalities. Their play bouts were also more interesting because they were much better coordinated and stronger than the younger ones.

Calf play takes several forms. Most often the calves engage in head-on sparring but since they have no tusks they merely run at each other and butt their heads together. They also chase one another, the pursuer often grabbing the tale of the fleer. They love climbing on top of each other, particularly when they are in a mud wallow. Then there is often a great heap of calves all wiggling and scrabbling and slipping and sliding in the glutinous mud. Sometimes calves will mount each other in the sexual posture and it is almost always a male calf who does the mounting.

There are definite differences in physical development and behavior between the sexes at a very early age. To skip ahead a little in time, in 1982, much to my delight, my former camp-mate Phyllis Lee returned to Amboseli after completing her Ph.D. on social development in infant vervet monkeys. With sponsorship from the National Geographic Society, she began an intensive two-year study of elephant calves. One of her main interests was these sex differences. Using both my 1980–81 study and her later research, we combined results and published a joint scientific paper on the differences in suckling behavior and growth rates of male and female calves and the differences in maternal behavior toward the sexes.

It is an accepted tenet of evolutionary biology that an animal's primary goal is reproductive success; that is, each animal attempts to have as many healthy, robust offspring as possible reach reproductive age and breed. In species with polygynous mating systems, a male might mate with many females, thereby producing far more offspring than a single female could in her lifetime. In these species, it is predicted that a mother should invest more time and energy in her male offspring than in her female offspring. In the case of elephants, if a female can raise one large, strong male who will father many calves, she will probably have more descendants than if she had several successful daughters. However, some males may never compete successfully and may end up siring no offspring, while daughters will probably produce some calves, so biologically it may be a gamble to invest more in sons. All the same, theoretically it is a chance worth taking. I hasten to add that the goal that motivates an individual's behavior is not a conscious one.

The red deer, an animal that has been extensively studied over many years by a team of scientists led by Tim Clutton-Brock, is similar to the elephant in that females live in small families, males compete intensively for females, and male size and strength are determining factors in who mates with the females. Clutton-Brock and his colleagues showed that male offspring of red deer received more milk and were weaned at an older age than female offspring. The early growth of the stag seemed to be very important in how successful it was as an adult.

Phyllis and I wondered if anything like this was going on among elephants. Our data revealed that males attempted to suckle more often than females and that they were also more successful in their attempts. When a male tried to suckle and the mother would not stand still or walked away after only a few seconds, a male calf tended to scream in protest more often than a female. Elephant mothers were remarkably tolerant of their calves' tantrums and always gave in and let them suckle. Thus, since the males demanded more milk, they actually got more. At the same time we found that males grew faster from birth, and presumably they needed more milk for this growth. Supporting this theory was the mortality rates of male and female calves. During wet years there was no difference in calf survivorship between males and females, but during drought years significantly more males died than females. We postulated that during dry years some mothers were unable to sustain milk production at a level that met the metabolic requirements of their sons and so male calves were more likely to die. In addition we found that if a mother had a surviving male calf the interval between the birth of that calf and the next calf was longer than if the first calf was female. Thus, since mothers suckled calves right up until the birth of their next calves, male calves tended to be weaned at an older age than females.

Therefore, our results show that mothers do invest more in male calves in the early years but we cannot conclude that they are doing so *in order* for males to grow faster and thus develop into large, strong adults with better chances at reproductive success. Mothers may be forced to invest more in males simply *because* they grow faster, demand more, and die if they do not get sufficient nutrition. The results in either case are the same in evolutionary terms. Nevertheless, in the long run maternal investment in the two sexes seems to even out for elephants, for sons leave their mothers at puberty and daughters stay on in the family, where they are supported and aided by their mothers for years to come.

During 1980 an event occurred that gave me an unexpected opportunity to look at sex differences between siblings. By the end of May, 99 calves had been born to the Amboseli population since the births had resumed in late November 1978. This had been a good time for the elephants in terms of food abundance, lack of poaching and spearing, and particularly reproductive success, and I decided that when the one hundredth calf was born I would have a small celebration. In early June I had to go to Nairobi to do some work for the African Wildlife Foundation. I asked Joyce to keep her eye out for Estella, the matriarch of the EA family, because I expected her to have the one hundredth calf. I had seen Estella in estrus 22 months before. Joyce came to Nairobi about a week after I did and said, "Boy, do I have news for you. Estella gave birth to the one hundredth calf but also to the one hundred and first. She had twins!"

I was thrilled and could not wait to see them. Twinning had been recorded in elephants before, but I had not seen it yet in Amboseli. In areas where elephants are culled and postmortems are carried out, twin fetuses are found from time to time, and recently there was a case of triplet fetuses in an elephant in Zimbabwe. When the well-known reproductive biologist Richard Laws was working in East Africa in the 1960s, he estimated the incidence of twinning in elephants to be about 1 percent.

My statistic of twins born on the one hundredth birth fit rather neatly and pleased Dr. Laws when I told him.

Although conceptions of twins occur at a rate similar to that in humans, what appears to be rare among elephants is for both twins to survive. Twins were born to a young female in Lake Manyara in 1976 but one died when it was five months old. There was also a case of twins born in the Aberdare National Park in Kenya, but again one died after a few months. Under most circumstances a female can probably not produce enough milk to nourish two calves. I was very anxious to see what would happen to Estella's twins.

I went back down to Amboseli on June 16 and found Estella two days later. There was a newborn calf with her, but only one. I waited, hoping the other might be lying down where I could not see it, but it was soon clear that it was not there. I was terribly disappointed, thinking that it had died and I had missed seeing them together. Then from around a corner came Elfrida, Estella's nine-year-old daughter, with another newborn calf. This was definitely not hers because she had no breast development at all. They joined Estella and the two calves stood together. Then I could see that they were almost exactly the same size and they had the same tight curly black hair on their foreheads. One was a male and one a female.

A while later the twins began to suckle from Estella with one on one side and the other on the other side. It started out peacefully enough but then the male calf ducked under Estella and pushed his sister off the nipple. She went around to the other side, and again he rushed under and pushed her off, with hoarse cries and growling rumbles from her or possibly from both of them. I was surprised at how much aggression there was between them. It was always the male, who was just slightly bigger, who initiated the aggression, and he was usually successful at disrupting the female's suckling.

Over the next week as I watched them I worried that the female would not make it, but she was smarter than I thought. When the twins were not suckling they did not fight and they actually spent quite a lot of time playing together. One day I discovered what the female was doing to gain equal suckling time. I watched the two of them playing for about ten minutes, which was a long play bout for small calves. They ran and fell in the long grass, climbed on one another, and butted each other head to head. Finally the male tired and lay down to go to sleep. As soon as he did the female moved to the far side of Estella where he could not see her and suckled in peace for over three minutes. Eventually, after several weeks, the fighting stopped and I saw no aggression between them again.

I added these two calves to my set of six newborns and watched them systematically over the next nine months. They were given codes (ES0 and ET0) but I also decided to give them names right away because they were special. I could not think of any famous twins whose names started with *E. Estella* means "star," so I decided on an astronomical theme and came up with Eclipse for the female (there was a total eclipse of the sun in Kenya in 1980) and Equinox for the male. Followers of astrological charts were pleased to know that the twins were born under the sign of Gemini.

I never stopped worrying about the twins. I wanted very much for them to survive and each time I found the EAs I held my breath until I saw that both were there. But if anyone was going to have twins that survived I thought it would be a female like Estella. She was a matriarch, an older, experienced mother with several surviving offspring, and it just so happened that her next oldest calves were both females: Elvira, who was five years old, and Elfrida, who was nine. They each spent a lot of time caring for the twins.

Although 1980 was not a year of good rainfall there was still abundant vegetation from the previous three wet years and Estella had plenty to eat and appeared able to produce enough milk for both calves. The twins survived and I expect it was a rare case of all the right circumstances coming together. Today, as I write this, the twins have reached their sixth birthday and they have a younger sister, born in December 1985. They were completely weaned after her birth. Equinox is bigger than his sister, Eclipse, but they are both considerably smaller than other calves of the same age. However, they seem in good condition. With cattle, when a cow gives birth to male and female twins, the female is always barren and is called a "freemartin." Many people have asked me if this will happen to the elephant twins. I will have to wait at least another six or seven years to find out.

In the meantime, the T families had two more births in 1980. In early November Tonie of the TA family gave birth near my camp. The calf was either born dead or it died shortly after. I suspect that stillbirths and neonatal deaths are fairly common in females giving birth to their first calves, as Tonie was doing. What was unusual with this birth and death was that Tonie refused to abandon the dead calf. I was in Nairobi, but fortunately once again Joyce was on the scene. In addition some local tourists from Nairobi, Clare and Brian Shorter, keen wildlife enthusiasts, also saw Tonie and the dead calf, and they took pictures and notes. Joyce had found Tonie when the calf was already dead. By then the rest of her family had left her to go off and feed. Tonie was out in the hot sun in a desolate area of hard pan and saltbushes. She was standing over the dead calf and threatening anyone or anything that tried to come near it. Joyce's observations indicated that Tonie remained with it for four full days.

The Shorters discovered Tonie on the third day. They arrived to find two lionesses dragging the dead baby away. Tonie suddenly saw what was happening and charged the lionesses, chasing them off. She then stood over the carcass, keeping the lions at bay for what the Shorters estimated was 30 hours. On the second morning after they had first sighted her, they found four lions feeding on the dead calf and no sign of Tonie. She had finally given up her vigil, probably because of thirst. She had been a fierce and brave mother and I hoped she would soon get pregnant again and have better luck with her next calf.

I was not anticipating any other births in the T families during the remaining months of 1980. Three of the young females, Tara, Tilly, and Theodora, came into estrus for the first time that year and so they were expected to have babies in 1982, and the fourth young female, Tess, did not show signs of being pregnant. In fact I was wondering if she was able to have a calf because she was already 17 years old and had not yet given birth. The other adult females all had small calves except for Teresia. In 1980 Teresia was about 58 years old. She had had her last calf nine years before, in 1971, at the age of 49. The reproductive studies carried out on culled elephants suggested that females reached the equivalent of menopause around 55, at which point they entered a sort of grandmother class and their experience and wisdom were used to lead the families. I thought that Teresia had probably arrived at that time in her life. Once again the Amboseli elephants had a surprise in store for me.

On December 20 I was out in the Olodo Are woodlands and found a large aggregation of elephants. The short rains had come by then and there was green grass and the elephants were feeling good. I had been in the section with the Ts, trying to do focal samples on the babies. The elephants had been walking and feeding when suddenly something sparked them off and they started running across the open pan, all 200 of them. But it was not panic running, which is quiet and fast and smooth. This was something I called "floppy running," where the elephants put their heads down and let their ears and trunk hang loose so that they flap and wave about as they run. At the same time they make the loud, pulsating play trumpet. My notes read:

1020 The large group behind starts to move fast but play trumpeting. They all start running. It has just started to rain a little. Now all running and a *great deal* of noise.

1030 Still a lot of noise. Floppy running.

1040 Still floppy running and trumpeting, whirling and attacking. How can one do a serious study of animals that behave this way!

1045 Back to other section. They come onto pan and all floppy run full speed across, play trumpeting like mad. Even Slit Ear and Big T [old matriarch of the BBs]. Except Teresia who sedately *walks* across. The Ts lift heads and charge at imaginary things.

On December 26 I found all the Ts together, resting under some trees. Teresia was on her own, 30 yards away, standing out in the sun. She looked tired, and definitely older than ever, partially because she had broken off her other tusk sometime during that year and only had two stumps left. She had seemed slow and lethargic on the day of the mass play, and today she was acting strangely by resting away from her family. I thought her behavior might be due to old age.

On New Year's Day, 1981, I was out in Longinye swamp and came upon the TCs and TDs. I did a census of the families as they came by and as usual Teresia was at the rear, this time with Theodora. But as they got closer I realized that there was also a tiny newborn baby with her. It was walking well and looked strong and healthy. It still had a bit of umbilical cord hanging down and, as with all newborn calves, the backs of its ears were bright pink. I thought it was probably a day or two old, making it born on the thirtieth or thirty-first of December, 1980. I had not noticed any change in Teresia's breasts, but perhaps I had not been looking carefully. She now had nearly full breasts and was definitely lactating. I thought it was remarkable that a female her age could still have a calf. No wonder she had been slow and sedate ten days before.

I stayed with the Ts for the next four hours, mostly doing samples on the other calves but keeping an eye on Teresia's as well. It was a male and seemed fine to me. The only unusual thing I noted about him was that he seemed to suckle more often than other newborns.

I saw Teresia and her calf two more times in the next two weeks and then on January 20 I found the family again. The calf was gone and Teresia was secreting from her temporal glands. For some reason I was not surprised. I think I half expected it not to survive. Earlier in the year Virginia of the VA family, whom I estimated to be 52 years old, gave birth to a calf that lived for only about a month. She too had not had a calf since 1971. It appeared that, rather than a complete cessation of ovulation, the rate at which females breed slows down as they get older, and if they do give birth the chances of the calf surviving are reduced. It could be that Teresia and Virginia could no longer produce enough milk for their calves or that the calves had genetic defects. I have no proof for either theory but for whatever reason these two calves did not survive during a time when calf mortality in general was very low. Neither female produced another calf after 1980, although both continued to come into estrus from time to time.

Teresia may have lost this final calf but on the whole she probably did fairly well over her lifetime. At the end of 1980 she remained with her son Tolstoy, a tall young male who looked as if he would turn into a large, successful bull, and her grandson Tim, who was also showing signs of one day becoming a fine bull. In Teresia's long life she had probably raised several other sons who had become independent before my study began and were very likely fathering her grandchildren. On the female side of the lineage she remained with her daughter Theodora, her granddaughter Tallulah, and her great grand-daughter Tao. Her genes were certainly carrying on into the next generations.

The Ts in general had experienced a very good year in 1980. They increased their numbers and at least three more females, possibly four, were pregnant. Shortly after Slit Ear and Tia gave birth to their new calves, I gave their '76 calves names. It was getting even more difficult to find unique three-letter beginnings to "T" names but I came up with Tabitha for Slit Ear's female and Taddeus for Tia's male. There were now 24 animals in the four families. After the death of Teresia's calf the structure of the four families looked like this:

TA	Tuskless (62)	♀
	Tuo (80)	♂
	Teddy (66)	♂
	Tonie (67)	♀
	Tilly (70)	♀
TB	Tania (44)	♀
	Tom (79)	♂
	Toby (73)	♂
TC	Slit Ear (36)	♀
	Slo (80)	♀
	Tabitha (76)	♀
	Tamar (71)	♀
	Tara (67)	♀
	Tess (63)	♀
	Tia (50)	♀
	Tio (80)	♂
	Taddeus (76)	♂
	Right Fang (70)	♂

TD	Teresia (22)	♀
	Tolstoy (71)	♂
	Theodora (67)	♀
	Tallulah (63)	♀
	Tao (80)	♀
	Tim (69)	♂

Chapter VII

Elephants and People

1981

It was November 1978, on a dark, moonless night. The sun had set two hours before, when Tuskless had led her little band out of Longinye swamp. They traveled first to the Amboseli Lodge garbage dump and then to the Safari Lodge pit, checking them out for new deposits. They found some fresh refuse in the pit and picked out what they liked before continuing on across the open pan to the Ol Tukai Orok woodlands and the public campsite. They wandered through the area, deliberately passing close to the camps and carefully smelling and searching for food boxes. Most of the campers had finished eating and had retired to their tents, but in one clearing some people were sitting around a table enjoying their dinner.

Tuskless broke through the palms at the edge of the glade and stood for a moment taking in the scene. The campers looked up and saw her, became excited, and started to comment and laugh. Tuskless moved toward them in a straight line. The laughter grew louder and then took on a hysterical note. The

177

people leapt out of their chairs and started shouting, clapping their hands, waving their arms, and banging pots and pans together. Tuskless ignored them and kept coming. The campers backed off and fled for their car. They huddled inside with all the doors and windows closed, watching helplessly as Tuskless, joined by her calf and Teddy, Tonie, and Tilly, ate every bit of food they could find on the table and in the safari boxes on the ground nearby. When all conceivable food items were devoured, the elephants ambled away into the palms. The campers waited a long time before cautiously emerging to inspect the damage.

In the meantime Tuskless and her family met up with Tania and her calf and exchanged deep greeting rumbles and ear flaps. They fed on grass and palm fronds for a while before emerging into a large clearing where a permanent camp had been based for a long time. They knew it well. They also knew the people who lived there—a tiny, fierce African man and others who got in their vehicles and frightened the elephants away when they came too close. Tuskless knew she had to be careful when she came into this camp and so she never just boldly walked in but used an indirect approach.

Tuskless, Teddy, and the others began feeding around the periphery of this camp, apparently minding their own business and concentrating on eating grass. However, surreptitiously, Tuskless turned the tip of her trunk toward the kitchen and the tents to smell what and who was there. Wafting from the place where the food was kept was the delicious odor of very ripe bananas. Tuskless loved bananas. She had first tasted them at the lodge feeding place, where a man came twice a day with a wheelbarrow full of vegetable and fruit scraps. These were usually peels and old, outer leaves of lettuces and cabbages. But one day there was a whole stalk of overripe, almost fermented bananas. Tuskless liked the smell immediately, plucked one off the stalk, popped it in her mouth, and savored the sweet taste. She managed to eat most of them, sharing as few as possible with the others. From that day on she particularly sought out bananas.

On this evening the smell was irresistible. She moved closer to the kitchen—a flimsy structure made out of a few wooden posts, sisal poles, chicken wire, and roofing felt. She could smell other good things as well—pineapples, oranges, and various vegetables. She also noticed that it was unusually quiet in the camp tonight. There were no voices and there was no strong, immediate smell of human. She stopped feeding and rumbled gently. Tania answered from within the palms with her unmistakable long, rolling rumble, but stayed where she was. Teddy, Tonie, and Tilly also answered and they walked over to join Tuskless.

These four animals, plus Tuskless's two-year-old calf, slowly approached the kitchen with their trunks held out in front of them. They got within a few feet and stopped and listened. Still there were no sounds of people. Tuskless reached her trunk out and felt the outer wall of the kitchen. She wrapped her trunk around a sisal pole and pulled. There was a tearing noise as the pole was wrenched away. No one came out of a tent to chase them. The others moved forward and also started pulling on the poles and the wire. Tuskless quickly made a sizable hole in one wall of the kitchen. She reached her trunk in and tried to feel for the bananas, but they were on the opposite side of the kitchen and a table and a cupboard were in her way. She became impatient and leaned her head against the wall and pushed. The building swayed before collapsing sideways. This did not help matters; the bananas were under the tilted-over wall and roof. But Tuskless was undeterred. She went around to the other side, which was more or less intact, and started forcing her way through that wall. Once she got her head inside she began to pull things out, push things over, open things up, and gather in as much food as she could in as short a time as possible.

The other family members had crowded in beside and behind Tuskless to get as much as they could too. The calf was doing very well because she could squeeze in under the fallen roof and get to things the others could not. Only Tania and her calf were uninvolved. They continued to feed on the natural vegetation at the periphery of the camp and refused to join the feast.

The bananas went first, of course, closely followed by the pineapples, oranges, mangoes, and papayas. These fruits, even large pineapples, could be placed whole in the mouth and crushed in the huge grinding molars with a great gush of juice.

It was near-ecstasy. The vegetables—carrots, potatoes, toma-toes, onions, lettuces, cabbages, cauliflowers, cucumbers, avo-cados, string beans, zucchini, eggplants—went soon after. Most of the fruits and vegetables were stored in tin trunks but this posed no problem for an elephant. The tin trunk was lifted up and turned upside down and stepped on a bit and kicked about if necessary, until it disgorged its contents.

Once the fresh produce was finished Tuskless began to search around for other delicacies. She could smell bread and cookies and crackers in one wooden cupboard. This was knocked down and smashed open. The bread was quickly snatched up and then the cookies and crackers were eaten, box or plastic bag and all. Next a small camping fridge and a cooler were crushed, but neither contained anything interesting. Another cupboard with glasses and plates in it was tipped sideways, and broken glass joined the other wreckage on the dirt floor of the kitchen. The bigger fridge was pulled and dragged from its place until it toppled over, disconnected from its gas cylinder. Noxious propane gas hissed out of the dangling tube with a sound and smell that would certainly have sent lesser elephants away.

Just as Tuskless and Teddy were tackling the third cupboard, with the jams and spices in it, they heard the engine of a car coming in their direction. A few minutes later the headlights swept around the clump of palm trees at the clearing entrance and shown brightly on the five elephants standing in, on, and around the totally destroyed kitchen. The car hesitated for only a moment and then the engine roared and the car headed for them. Each elephant grabbed a last trunkful and started backing or turning away, reluctantly abandoning the banquet. Tuskless held her ground a little longer than the others but the vehicle came straight for her, making loud engine noises, and she too left, carrying a box of spaghetti in her trunk while chewing on a paper bag full of half a pound of garlic.

I can still close my eyes and vividly see that disastrous scene as I drove around the corner into the camp and our lights shown on what had been the kitchen. It was November 4, 1978, two to three years before most of the other events in this chapter, but for me it was the beginning of a new attitude toward some of the elephants and a new way of living in the camp with them.

On this evening I was with my camp-mate Phyllis and Warren and Genny Garst, filmmakers from the *Wild Kingdom* TV program, who were staying with us to film Phyllis's work on the vervets. (They had done a program on the elephant project the year before.) When I saw the elephants I had no option but to go straight for them. They knew my Land-Rover and knew I would harass them till they left. I would never actually hit an elephant with my vehicle but I would get within a few feet, and this was enough to make one move.

As soon as the elephants left, I quickly got out of the Land-Rover and crawled into the kitchen because I could hear and smell gas escaping. I found the gas cylinder and removed the regulator to close off the valve. I thought later how horrible it would have been if a spark had set off the gas while the elephants were there. The rickety, dry kitchen would have gone up in flames in seconds.

As angry as I was at Tuskless and her gang, I did not want them to get killed or wounded. And in any case I mostly blamed myself for what had happened that night. An unusual set of circumstances had conspired to lead Tuskless to our kitchen. First and probably most important, my cook/camp worker, Masaku, went away for five days' leave on November 3. On the same day Phyllis came back from a trip to Nairobi with a huge load of groceries, meat, and fresh fruit and vegetables, including some overripe bananas with which she was going to make banana bread. The next day the Garsts arrived with yet another huge load of food. Since it was their first night there and since Masaku was away, they invited us to have dinner at one of the lodges. We gladly accepted.

Previously we had left the camp unattended on many occasions. Elephants had broken into the kitchen two years before when no one had been in the camp. It may or may not have been Tuskless; there were also a few males around who raided camps then. They had bashed through one corner and had dragged out a tin trunk, crumpled it fairly badly, and eaten its contents. The trunk had been standing open. After that we carefully put food away and closed all the trunks to reduce the tempting odors, and we had not been raided again. On the evening of the fourth we put away as much food as we could, but there was so much that we could barely close the trunks. We even locked some of the Garsts' food in their Toyota so that it would not be left exposed in the kitchen. In the early evening we had seen Tuskless leaving Longinye and heading in our general direction. We should have known better.

By the latter half of 1978 Tuskless had become very, very bad about raiding camps. She and her group sometimes raided as many as three or four camps a day, both at night and in the daytime. She checked out our camp almost daily, but when she got too interested in the kitchen she was chased away by Masaku, who, despite having only one eye, was wickedly accurate at throwing empty wine bottles. Although the bottles did not hurt, Tuskless did not appreciate having one land between her eyes. Masaku usually just had to come out of his tent and she and the others would move off, nonchalantly feeding as if they had never even known there was a kitchen in camp. We thought we had arrived at some sort of understanding with Tuskless but obviously had not. She was just waiting for her chance.

And what a reward she got. She and the others ate all of our fresh food, except for what was in the Toyota, and most of our staples. What they did not eat they mostly ruined. It was a horrific mess. We started cleaning things up immediately, working until after midnight. We ended up putting the refrigerator and one cupboard with the remaining food in a closed vehicle in case the elephants decided to come back.

The next morning we woke to the full glory of the wreckage. It definitely looked worse in daylight. We began sifting and sorting through the debris. I think Phyllis and I were most annoyed to discover that they had broken the full bottle of Scotch that Phyllis had just bought for the camp. It was a luxury we sometimes rewarded ourselves with but one we could rarely afford. I was also unhappy about my crushed camping fridge, which I had had for several years and was fond of. My new cooler was unrecognizable.

Warren and Genny pitched in and helped us clean up and repair things. After a preliminary sorting out Phyllis and I went over to the park headquarters at Ol Tukai and picked up three *fundis* (Kiswahili for skilled workers, in this case carpenters) and brought them back to rebuild the kitchen. They tore everything down and started from scratch. By working extremely hard, we managed to get the kitchen finished by dark. There was not much food in it, but at least we could set up the refrigerator, put what there was away, and cook something for dinner.

From that day on we never left the camp unguarded for more than the ten minutes it took to drive to Ol Tukai and back on Sundays, when we dropped Masaku off in the morning or picked him up again in the evening. Anytime Masaku was away one of us would have to stay in to guard the camp. Even now Tuskless and her group come almost daily to feed in and around the camp, checking things out and surreptitiously smelling in the direction of the kitchen. We know if we relax our guard that she will be in there in a flash.

The story behind Tuskless's camp raiding and garbage eating is the old one of people trying to become too familiar with wild animals. It is similar to the problems with bears in the U.S. national parks. Ironically, two months before our raid, I was asked by the warden at the time, Joe Kioko, to write a report on elephant raiding, including suggestions for alleviating the problem. Part of my report follows:

It is necessary to understand the background to the problem in order to have any hope of solving it. I believe that the whole problem lies in the improper disposal of garbage, aggravated by a period of actual feeding of the elephants at Amboseli Lodge. The garbage pits at the Amboseli Lodge and the Ol Tukai bandas [small cabins with self-help accommodation] are merely slight depressions in the ground. The TA family is one of the central Amboseli groups that use the Longinye swamp in the day and the Ol Tukai Orok woodlands in the morning and evening. Their route to Longinye in the morning passes near the Amboseli Lodge garbage pit, and their route out of the swamp at night passes near the bandas' garbage pit. At some point, I estimate around 1973, one or more of the elephants in this family discovered that the smells coming from these pits actually came from foods that they found highly palatable. They then went back to the same spot each time they passed that way to try to find more of the new foods. For a year or so they did not appear to make special trips to the garbage but checked it out when they were in the area. Then in 1975 the Amboseli Lodge built an elephant feeding station just beyond the swimming pool and delivered wheelbarrows full of vegetable and fruit refuse from the kitchen. The TA and TB families, and even some other families, took to the feeding station almost immediately. The other groups visited it if they happened to be going that way, but the Ts soon began to make several trips to the feeding station each day, actually moving out of Longinye in the middle of the day to go to the Lodge. By 1976 the TA family was very familiar with the smells of human food and started being attracted to the same odors in the campsites.

At the end of 1976 the matriarch of the TA family, Torn Ear, was killed by poachers outside the Park, and the loss of her leadership allowed her daughter Tuskless free rein in determining the movements of her group. Tania of the closely bonded TB family did not appear to influence Tuskless even though she was much older. With the loss of Torn Ear the situation went from bad to worse. Tuskless and Teddy started raiding camps far more frequently. The family did leave the Park for the wet season migrations in 1977 but was back at the Lodge as soon as the dry season began.

Later in 1977 the feeding of the elephants at the Lodge was stopped at the order of the Ministry, but, unfortunately, the garbage from the kitchens continued to be placed in a shallow depression which the elephants had no trouble reaching. They simply concentrated on the garbage pits and camps once the feeding station was taken down.

I made various recommendations, the first and most important being that the garbage areas be made elephant-proof with deep vertical-sided pits and that the refuse in these pits be burned every day. I also advised trying to educate the campers, telling them about the situation and asking them to help by locking all their food away in vehicles at night. I further suggested that once the garbage was out of reach, the elephants in the campsites be chased away by the rangers with thunder flashes. I concluded:

The most important thing is that when the elephants return to the Park [after the next wet-season migrations] there be no garbage accessible to them. If they get even a small amount of garbage it will act as a positive reward for coming back and searching in that area. The positive reward is the key to how an animal searches for food and moves through its habitat. I believe that Tuskless and Teddy can reform with strong negative conditioning, that is, creating a situation where they associate bad experiences with trying to raid a camp. An important thing to remember is that these elephants do not need garbage to survive. An elephant eats approximately 300 lbs. of food each day. I estimate that Tuskless, who gets the most, does not eat more than 20 lbs. of garbage per day on a good day. This is less than 10% of her diet. There is no necessity to eat garbage, but they find it rich and palatable and so they go out of their way to seek it out.

I do not recommend killing Tuskless until these other measures have been tried first. In any case the death of Tuskless will not solve the basic problem. Other elephants from the same family may continue to raid, and inevitably other families will learn about garbage as well, and then members of those families will have to be shot. In the meantime the chance of a tourist being killed grows greater each day.

My report was accepted with appreciation and at least one of the measures was taken. Signs were put up in the campsites telling people to lock their food away in their vehicles, and the situation did improve for a while. However, the garbage situation, if anything, became worse. One problem for the parks administrators was that Ol Tukai, the area that encompassed the lodges and the self-help bandas, was not actual parks land but belonged to the local county council. Therefore, the parks personnel could only suggest that something be done about the garbage but they could not implement any changes themselves.

By 1981 the problem with the elephants was acute. Whereas once the lodges were trying to lure elephants in closer for the tourists to see, they were now distressed by the damage the elephants were doing. The elephants fed on the trees and plants around the lodge; they continually broke through a fence and got into the kitchen compound; and one dark night an extremely drunk lodge employee who was walking back from the canteen lurched straight into an elephant and got tusked through the side. Miraculously the elephant did not try to finish him off and the tusk missed any vital parts. He was back at work within two months.

I still do not know who tusked the man; possibly Tania or a more recent member of the garbage brigade, M76; obviously it was not Tuskless. What I find amazing is that no one had been hurt earlier. Over the years Tuskless and the others had had every opportunity to injure or kill tourists and lodge and parks employees, not to speak of researchers. And they had every provocation. Every day in the camps and around the bandas and lodges these elephants were shouted and run at, had things thrown at them, even had people come right up and hit them with sticks or brooms. All the elephant had to do was swing its trunk and the person would be dead. Elephants are incomprehensibly strong. In and around Amboseli they kill Maasai cows from time to time. They do not do it by knocking a cow over and tusking it; they just kill the cow (an animal weighing up to 1,000 pounds) with one swipe of the trunk. Think what that would do to a 150-pound man.

And yet during every single incident I have witnessed (and over those years there were many), Tuskless and the others went out of their way to avoid hurting a person. At times it was actually difficult not to step on or run over someone but they always swerved or backed quickly to avoid doing so. Sometimes these encounters could be amusing, although only after they were over. At the lodge with the feeding station, the manager had ordered a single line of painted white rocks placed along the periphery of the lawn. The feeding station was only about fifteen yards beyond. A sign was put up forbidding anyone to walk across the line. So being good, obedient tourists, as soon as the elephants came they picked up their cameras and ran out as far as the rocks. For some reason that I find difficult to fathom they thought they were perfectly safe behind the rocks, which were no more than eight inches high. The elephants sometimes found the tourists irritating, and Tania showed definite signs that she disapproved of the people standing there shouting and laughing. She would often give a sharp toss of her head or a short charge, which would send the tourists running and screaming toward the lodge. Her son Taabu was the worst offender. Being a young male on the verge of independence, he was full of high spirits and mischief. He would come walking over from the garbage pit to the feeding station with a particular jaunty gait that I recognized meant trouble. He would often carry a plastic or burlap bag he had found in the pit and then stand in front of his audience and whip the bag back and forth, up and down, over his head, onto his back, in between his legs, before finally flinging it in the air. Sometimes he would put his head down and charge the people; and lo and behold, he could actually run over the rocks.

One day Tania lost her temper and came charging at a hapless woman tourist who ran back toward the lodge, but fell halfway across the lawn. Tania, only a few feet behind her, skidded to a halt and towered above her. She backed up, turned, and ambled across the rocks and rejoined her family. She could have easily killed the woman, but for whatever reason she did not want to and had to use a substantial amount of energy to avoid doing so. (She left deep skid marks in the ground.) Every day this family had encounters with people and yet with the exception of the drunken room steward no one had been touched.

The lodge management finally realized that they had a serious problem on their hands. After they took down the feeding station, leaving only a water hole out beyond the line of rocks, an independent businessman was brought in to build an "elephant-proof" fence around the garbage pit. First he built one of railway ties, stout posts, and thick cable. The elephants literally walked through it. He tried a different design and they walked through that as well. Finally he got thicker posts, thicker cable, and cemented the heavy railway ties deep into the ground. He was positive that this was an elephant-proof fence. The day came to close it up. The lodge manager and other staff came to watch, and the fence builder had his camera ready. Tuskless and Tania and the others arrived at their usual time to feed at the garbage pit. They stopped when they saw the fence, reached their trunks out, smelling, stood for a bit, and then Tuskless went forward. She lifted her foot, placed it on a taut cable, pressed down, and a corner post came right out of the ground. She pushed the fence over and walked in. The others quickly joined her. Tuskless was only in her late teens at the time, and probably weighed no more than 5,000 pounds. It is wonderful to think what a big adult male weighing 12,000 pounds could have done. The businessman gave up for a while after that. Eventually he tried electric fencing, which proved to be far more successful.

I cannot help but love and admire Tuskless. She is smart and brave, inventive and gutsy, and at the same time she has to be one of the sweetest-natured animals I have ever known. I can never become deeply angry with her, regardless of how badly she and the others sometimes behave. I enjoy having them right in the camp, even right next to my tent, as long as they are feeding on grass or palms. The TAs and TBs over the years have become our camp companions. They are still wild animals but we accept each other and we do have some understanding of what is allowed and what is not.

Fortunately, for the elephants, for my study, for the parks authorities and the lodge staff, no other family units became garbage raiders. The TCs and TDs and the EAs and EBs sometimes went to the feeding station, but once that was taken down they did not seek the garbage pits or the camps. All the other elephants in the population fed on natural vegetation and moved through the park according to normal foraging patterns. Only two males from other families, M76 and M82, turned to garbage. M82 was placid but M76 could be very nasty at times and very persistent.

It was M76 who began to try to open up cars to get at the food inside. Irate campers would arrive at Warden Bob Oguya's (Bob replaced Joe Kioko in 1978) office door in the morning to show him crumpled car trunks, dented doors, and broken windows. Bob dispatched rangers to the campsites and they fired over the heads of the elephants or threw thunder flashes at them, but this had virtually no effect. After a while even I thought it was time to consider shooting M76. He was very different from Tuskless and I thought he would hurt or kill someone.

A solution to the campsite problem came in a roundabout way. When Amboseli had been made into a national park, it was agreed that the public campsite should be placed just outside the park on a Maasai group ranch. In this way the Maasai could reap some of the benefits of tourism by collecting the camping fees. However, the Maasai elders, who were very conservative people, were suspicious of this plan, thinking it was a way for the park to take more of their land away. It took the younger, educated Maasai officers of the group ranch a long time to convince them the ranch would make money from the scheme. The elders finally agreed; and after another long delay in deciding where it should be (the elders did not want it where there were any trees because then they would not be able to bring their goats in to feed on the acacia pods), they cleared an area in the *Acacia tortilis* woodland on the southwest border of the park, built some latrines and a small house for an attendant, and in June 1982 declared the public campsite open.

The parks authorities very kindly allowed my research camp to stay where it was, but all the public campsites near us were closed down. With no easy prey around, I was a little worried that Tuskless might make a concerted effort to raid my camp, but this did not occur. She did not move to the new campsite either. It was way out of her dry-season home range and in any case she still had the garbage pits near the lodges. Thus the campsite problem was solved, but the behavior of the TAs, TBs, M76, and M82 in Ol Tukai continued to be troublesome.

In the meantime my camp was left in relative peace. It was delightfully quiet and calm without the other campers nearby. Even when the campers were there, it had always been an oasis in dry, dusty Amboseli, a cool and relaxing place to come back to after a hot day in the field. Next to the elephants themselves, the camp is the thing that makes me most happy in the work I do.

I chose the place for the camp in August 1975. The area was well behind the other campsites. To enter it one took a circuitous route through a stand of *Acacia xanthophloea* trees and around clumps of *Phoenix reclinata* palms. Coming around the last clump of palms one emerged into a large clearing covered in thick *Cynodon dactylon* grass, fringed on the western and eastern sides by more palms and acacia trees. In the center of the glade, three huge, old, yellow-barked acacia trees shaded nearly the whole area. To the north was another glade, also ringed with acacias and palms, and to the south was a long green meadow at the bottom of which were more large trees and above which rose Kilimanjaro in all her majesty. It was, and continues to be after eleven years, a sight to make my heart soar.

When I received a grant from the African Wildlife Foundation in 1975 to continue my study, I considered the various possibilities for accommodation in Amboseli. During the part-time study I had stayed at either Jonah's house or the Ol Tukai bandas. Jonah's house was not available because he was going to be spending more time in Amboseli himself; and the bandas did not appeal to me because they were in a busy, populated area. I wanted to set up a permanent camp in the park. I had already spent some time on my own in Amboseli and was not worried about living by myself, despite the fact that another woman researcher, Melanie Fuller, had been killed (probably by a lion; it was never satisfactorily determined) in Amboseli two years before. In fact, in 1974 I experimented with loneliness and with the possible "ghosts" of my friend Melanie by spending two weeks at Jonah's house with no one else in the park I even knew. It *was* lonely and I could not get out to work often because it was the height of the rains and I could not drive off, or even on, most of the roads in my little Renault 4. The death of Melanie, who had also stayed in Jonah's house, did not haunt me or deter me, although I found it very disturbing and sad.

I spent that time in Amboseli working on the book I was writing and on the elephant recognition files and family structures. I fed crumbs to the birds and talked to them; and one night after I had been there for ten days without speaking to anyone except the gas pump attendant, I had a vivid dream that an open car drove up to the house, and sitting in the passenger seat was a wooden, talking dummy. I was immensely pleased and completely satisfied to be getting a dummy. It seemed to be all I wanted.

I do not know if the dream proved I could live by myself or not. In any case, as it turned out, I have rarely lived in Amboseli on my own. At the same time that I wanted to set up a camp in

Amboseli, a Ph.D. student from Colorado State University, Wesley Henry, was starting a study on tourist impact and behavior. Wes was a tall, bespectacled, very serious young man who seemed good-natured and competent in the bush. I am not sure what he thought about me, but we decided to join forces and set up a camp together. He liked the site I had chosen and on September 12, 1975, we pitched our tents.

The camp evolved over the years. We started out with two tents: Wes's big (10×16) two-room canvas Eureka tent and my medium-sized (10×10) locally made canvas tent. We put Wes's tent under the three big trees in the middle of the glade. Wes used one of the rooms of his tent as a bedroom/study area and the other side was used jointly by us as a kitchen and dining room. My tent, placed under another tree at the southern end of our clearing facing the mountain, was my combined bedroom and study. It contained a small metal camp bed, a table and chair used as a desk, a cupboard for clothes, and various wooden boxes for storing things. Wes's vehicle was a VW camper in which he also stored his clothes and research materials.

Wes contributed a refrigerator from his camper and I contributed a two-burner stove (both ran off a gas cylinder), a round dining table with four traditional canvas and wooden camp chairs, and a large sisal carpet. Wes turned out to be quite a good cook, and he also liked building things and working around the camp. His first project was a shower, which was ingeniously designed. A 44-gallon drum was placed over a fire to heat the water. Into the drum were welded a pipe with a faucet and a pressure valve. A hose was attached to the pipe and run up to a shower bucket, which was suspended from a wooden frame built over a large hole in the ground covered with wooden planking. Once the water was heated by firewood, the pressure built up in the drum. Further pressure was added through the valve with a foot pump. When the pressure reached a certain point the hot water would rise through the hose and into the shower bucket. In the early days we just had a bit of burlap around a frame for privacy. Later we built a more substantial shower stall with sisal poles.

There was already a "long drop" latrine in the camp, nicely placed far away from the tents in a clump of palms. However, it was in a terrible state of disrepair and basically consisted of a large, deep hole with a few rotten wooden beams precariously suspended across the top. The superstructure had fallen down years before. Using the *choo,* which is the Kiswahili word for toilet, was probably the most adventurous thing that we had to undertake in that first month. Attempting to balance on a crumbling beam while holding a few palm fronds in front of one for privacy and trying not to think of the consequences of the beam breaking was a feature of the camp I was happy to dispense with as soon as possible. The rebuilding of the choo was one of our first priorities and we sent to Ol Tukai for some fundis. They erected a new outhouse and strengthened the floor but we were still left with the old beams, and it was not until 1978 that we put in a cement floor.

We had not done so earlier because, being in the park, we were not supposed to build any permanent structures. The camp was to remain easily dismantled and moved without leaving anything behind. We totally agreed with this policy and always tried to comply with it. The advent, several years later, of the cement floor in the choo came about because one day the warden in charge of works came to have tea with Phyllis and me and at some point asked to use our choo. He came back looking a bit shaken and said it was dangerous using the choo with beams like that. The very next day he sent his men over with a bag of cement. After that we had a beautiful choo, which we were inordinately proud of and which we could send our guests to without worrying whether they were going to come back.

Wes and I did all our own cooking, washing, and cleaning the first month, but we always planned to get a camp worker/cook to guard the camp and to do most of the time-consuming chores, leaving us free for our field research. Old East Africa hands have various theories about how to find a reliable employee. Some feel that it is best to bring in from the outside someone who has good references; others feel that it is better to ask someone working in the same place to recommend a friend or relative. I tend toward the latter theory, because I think it is nice for the new person to have at least one friend in the area and I reckon that someone with a good job is not going to jeopardize that job by recommending a person he cannot trust. I went to see the warden of Amboseli, Joe Kioko, and his

American-born wife, Christina. They had moved to Amboseli the year before and had become good friends of mine. They were always warm, hospitable, and extremely helpful and contributed significantly to making my first years in Amboseli happy and secure.

They had a very pleasant older man working for them, Maundu, who was of the Kamba tribe, which traditionally provided most of the gun bearers, trackers, cooks, and camp workers for hunting safari companies. We asked him if he knew of anyone who could cook and clean and wash and iron clothes, and, most important, was willing to live in the bush surrounded by large, dangerous wild animals. He said there was someone near his home who had worked as a camp cook and was presently out of a job. I gave Maundu some money to go home that weekend to see if he could bring the man back. His home was about 75 miles away, near the intriguingly named town of Sultan Hamud.

On Monday Wes and I stopped by the Kiokos house to see if Maundu had had any success. He was somewhat disappointed because the man he had had in mind had only a few days before found work and was off on a safari. He told me that he had brought back another man who had worked as a cook some years before. Did we want to meet him? We said all right and he went off to his quarters and a few minutes later came back followed by a tiny, decidedly evil-looking old man who glowered at us with only one eye. We exchanged the usual formal greetings and then he handed us an ancient piece of paper with yellow edges. It was a letter of "recommendation" from a previous employer dated 1954! According to the letter Masaku Sila had worked as a sort of kitchen *toto*, or assistant to the main-house servant/cook, for what sounded like a very colonial *memsahib* in Mombasa. It was not much of a recommendation because the woman, and I could almost hear her strident voice, wrote something like "This boy is trustworthy but needs supervision and does not work well on his own."

Masaku explained that after working for the family in Mombasa he was employed by the Kenya Railways as a manual laborer for 20 years before retiring. He did not like retirement and wanted to go back to work. At the time I met him he was probably about 55 years old. He did not know the date of his birth but thought it was around 1921. Wes and I went off and conferred with Christina. Masaku was not exactly what we had in mind. He certainly had no experience in the bush, he probably did not know how to cook, and he was fairly scary-looking, but he was there and Maundu was recommending him. We decided to hire him on a trial basis.

That was the beginning of a long, intense, sometimes joyful, often turbulent, always caring relationship between Masaku and me. Masaku took to camp life immediately. He handled the wild-animal visitors with great aplomb and apparently very little fear, but with enough respect not to get hurt. He was a natural cook. He remembered some things from his days in Mombasa, particularly fish and chips, which he made better than anyone else I knew. Wes and I both taught him other recipes, but by the second time he made a dish it was *always* better than the original dish that we had shown him. We bought an ammunition tin for use as an oven, and a cook from another research camp showed him only once how to make bread. This man's bread was hard and dense, but he was the only one around who could teach Masaku how to bake bread under camp conditions. From that day on Masaku made delicious, light fluffy bread, never anything like the other cook's. Soon we were having rich stews, piquant spaghetti sauces, spicy chilis, and hot curries cooked in a pressure cooker; roast chickens, rolled beef, and legs of lamb with garlic and rosemary cooked in the ammunition tin; plus simpler fare such as steaks and chops with various kinds of vegetables and potatoes.

We got permission from Joe Kioko to build a kitchen, assuring him that it could be taken down fairly easily. Masaku and Wes built it in a matter of days. As I mentioned earlier, the kitchen consisted of strong corner beams, sisal poles, chicken wire, and roofing felt, with a door made of wooden planks. The floor was hard-packed dirt. It just so happened that I had lived on a sisal plantation before moving to Amboseli. When a sisal plant flowers it sends up a long pole from which the flowers sprout. Afterward the whole plant dies and the pole dries and hardens. The poles vary in height and diameter but an average pole is about ten feet long and has a diameter of about three inches. I learned that they could be used for all sorts of things.

Masaku was also a whirlwind of activity around the camp:

sweeping out and dusting the tents, making beds, chopping firewood, heating water for our showers, and washing and ironing clothes, which he did very well with a charcoal iron (a skill to this day I find amazing; I cannot understand why the clothes do not get scorched by the iron and burned by sparks).

Masaku was never a shy, shadowy figure working quietly in the background. He was always very much a presence in the camp and his whole personality put a particular stamp on the place. He worked amazingly hard, and contrary to what his former employer wrote, he needed virtually no supervision. However, Masaku did need special care and attention. It was not obvious at first, in fact not even in the first year or two, but Masaku was subject to radical shifts of mood. I am not a psychiatrist or psychologist but from what I have read I would venture to guess that Masaku was a classic manic-depressive. During his highs he would dance and skip around the camp and sing old Kamba folk songs in a high, beautiful falsetto. He would clean everything with frantic intensity and want to start new projects such as painting the posts for the clothesline, which we preferred unpainted. Sometimes when we were out in the field, the new projects would get done before we had time to stop him.

In his lows, he was very quiet and his face changed completely, becoming blacker, with all lines turned down. He could barely speak and when he did it was usually mumbling to himself in the kitchen. Nothing was right during these times: The firewood was the wrong kind, the potatoes we brought from Nairobi were watery, the fish came in pieces that were so small that he could not even cook them, his joints ached, and he could not lift things. He often had to retire to his tent and rest. He was inconsolable and there was little we could do. During the lows the camp took on an atmosphere of gloom and we found ourselves practically whispering and tiptoeing around.

I learned how to deal with Masaku's highs and lows but sometimes they got to me and I would lose my patience (occasionally I thought the highs were worse than the lows); yet over the years we grew to love and respect each other. It was not as easy for the other people who later lived in the camp. As far as Masaku was concerned, Wes and I could do no wrong and every subsequent person who came to live in the camp never quite measured up to us. Some people handled it better than others. I sympathized with all of them. It must have been annoying in the extreme to be told every time one tried to do something in the camp, "But Cynthia and Wes didn't do it that way." It did not help that Masaku spoke only Kiswahili and most new people were just learning the language. But Masaku at his best was superb—smart, canny, funny, hardworking, totally reliable, totally trustworthy, loyal to me in a way that I will probably never experience again, and basically a kind and good man. At least some of these qualities were apparent to everyone who lived there, but Masaku could definitely drive some people to distraction.

Wes stayed in the camp for two years, leaving in July 1977 to complete his Ph.D. He eventually joined the U.S. Department of the Interior, but has come back to Kenya on short visits from time to time. (Once someone has lived in Africa it is hard to leave for good.) I stayed on my own for several months and briefly had a young assistant who lived at the camp with me, Joan Riback. Joan's most important legacy was to teach me backgammon. We had many a good game after dinner by candlelight. Even on my own I was not lonely in Amboseli. There was a community of researchers, and parks, lodge, and county-council employees. Long-term baboon and vervet projects each had at least one researcher based in Amboseli. The baboon project had a permanent house in Ol Tukai and the vervet researchers rented a banda. Jonah himself came from time to time and usually had a student working for him and living at his house. There were also researchers studying archaeology and paleontology who either rented bandas or camped. Two University of Nairobi students doing their master's degrees camped next to us and were excellent neighbors. At times there were over ten people in the field. The researchers, together with lodge managers, wardens, and others working in the park, had dinner at each other's camps and houses, met for drinks at the lodges, organized going-away parties for people leaving, and generally helped each other whenever possible.

In 1978, when I was still alone in the camp, Phyllis joined the vervet project, planning to rent a banda for the duration of her study. However, she was not happy there and I invited her to stay with me. We shared the camp very happily for two years. When Phyllis left in 1980, Joyce, who had been working in Amboseli for short periods since 1976, returned to start her Ph.D. research on the elephant bulls. Shortly after, Robert Seyfarth and Dorothy Cheney, a married couple who are the directors of the vervet project, decided to give up the bandas and move to the camp as well. A few months later a good friend of Joyce's, Cynthia Jensen, a plant ecologist, came to work with Jonah and also put up a tent in the camp. Robert and Dorothy (who returned in subsequent years) stayed for about a year until their new student, Sandy Andelman, arrived in 1981. Thus, in 1981, Joyce (elephants), Cynthia (grasslands), Sandy (vervets), and I lived in the camp. Having two Cynthias did not cause too many problems. For some reason known only to Masaku, he called Cynthia Jensen "Sin Sing," so there was no confusion for him. From 1978 on we had a second person working with Masaku, first his nephew, who had the charming and apt name of Boniface, and later Masaku's brother, Mwanga. The camp was busy, productive, and a very pleasant place to live.

In January 1981 I finished the focal sampling on the calves I had been following and planned to get back to analysis and writing. I did not intend to spend as much time in Amboseli as I had previously. In November 1980, I, along with two other friends, had rented a house in Nairobi where I hoped to work without the distraction of the elephants. In Amboseli I found it very hard to stick to my desk when I knew elephants were out there doing interesting things. I was reluctant to give up my time there, but in January 1981 an event occurred that forced the end of a particular era for me in Amboseli.

I had a cat in Amboseli, a sweet but rather eccentric gray male named "Moshi," which is Kiswahili for smoke. When I was about to move to Amboseli I tried to find a home for him with no success and so ended up, with much trepidation, taking him to Amboseli. I thought he might run away or get eaten or trampled by any one of the many animals that lived in and around the camp. It is theoretically against park regulations to have a domestic animal in a national park, but Amboseli was not run as a park when I first went there. Aside from several thousand cattle, sheep, and goats, there were 200 to 300 hundred cats in Ol Tukai around the two lodges and bandas, and some wandered as far as the campsites. I did not feel that one more cat was going to make a great deal of difference and I planned to keep him there only until the other cats were removed. (The cats were never totally eliminated and eventually the parks gave permission for some cats to remain in order to keep the rat population down.) In any case, my cat was a neutered tom who could not mate with any wild cats, he had had all his shots, and what was more he was a hopeless hunter so he was not going to have any effect on the bird population.

However, Moshi adapted to camp life with remarkable ease and was happier in Amboseli than in any other place he had ever lived. He stayed in the camp and, being a cat, slept most of the time anyway. He had his little routines, one of which was "hunting" every evening between 6:00 and 7:30 P.M. Moshi's idea of hunting was to sit beside the tree stump next to my tent and hope a mouse would jump into his mouth. In the five years he was there I think he caught only two mice. One year we had a literal plague of rats. It was a year of high rainfall and the rat population exploded. The Maasai began to contract bubonic plague and 20 people died before it was brought under control by preventive drugs. Rats were everywhere. Whenever I stopped the Land-Rover I could hear them rustling about. On some days Masaku caught seven or eight rats in the kitchen, sometimes two in the one trap. Moshi ate them all and soon became so conditioned that as soon as he heard the trap spring shut he would run to the kitchen.

Moshi seemed fairly oblivious to the dangers of living with wild animals. He had a few close encounters, one with a huge male baboon that got into my tent. There was a tremendous commotion in the tent and Masaku and I ran over to find the baboon leaving. Moshi was clearly frightened. I think the baboon had gone into the tent to look for food but found Moshi instead. Baboons do kill and eat small mammals and I think this one would have killed Moshi. Fortunately, he never came back. On another occasion a very large black mamba, over ten feet long, came sliding around the base of my tent and was about to come to the open entrance when I threw my shoe at the side of the tent. Moshi leapt off the bed to investigate. I grabbed him just as he was going out. I backed to the far side of the tent, waited a minute or so, went back to the front, looked out, and there was the mamba with its head raised waiting to strike. That is what I do not like about mambas—they refuse to run away like any other law-abiding snake. I called for Masaku and Boniface and they came running with sticks, and the snake slithered off into the palms. To this day I still shudder at the thought of what would have happened if the mamba had come into the tent or if Moshi had managed to get out before I caught him.

In the end it was a leopard that ended Moshi's stay in the camp. I had been thinking about taking Moshi to Nairobi once I began renting the house, although I knew he would be happier in Amboseli. I decided to delay making a decision until after I went to the States in February 1981 for a meeting of the New York Zoological Society. I was in Amboseli in January, finishing up my sampling before I left. A few days before I arrived for this stay Cynthia Jensen had discovered a leopard in the camp just after dark. The leopard had seen Moshi in his usual hunting spot next to my tent and had begun to stalk him when it was disturbed by Cynthia and a visitor. I was not happy to hear this story when I got there because leopards are notorious killers of domestic dogs and cats. I was worried that once the leopard knew Moshi was there it would keep coming back to try to get him. Unfortunately, I was right.

The night I arrived, January 14, I went to bed as usual with the tent door firmly zipped down and Moshi inside with me. Sometime in the middle of the night I woke up suddenly because Moshi had leapt off the bed with a loud thump. I sat up to see a leopard moving from the window right next to my bed around to the front of the tent. It was bright moonlight, and because I could see what the leopard was doing I was not afraid. I said something like "Go away!" It saw me and quickly slunk down the path to the choo. I went back to sleep almost immediately and woke as usual at dawn. It was not until I sat up in bed that I noticed that the window next to my bed had been ripped wide open. I think what must have happened is that the leopard came up to the tent looking for the cat; and Moshi, seeing the leopard, must have sat up or even moved to the window. At this point the leopard took a swipe at him and ripped through the screening, and Moshi leapt off the bed. The sound of the screen ripping must have woken me but I only registered the sound of Moshi jumping to the floor.

The ripped window forced my decision. I stayed in Amboseli for another week, keeping very careful watch over Moshi in the evenings and not sleeping all that soundly at night, and then took him with me to Nairobi. As I write this today Moshi is close to 16 years old, in excellent health, if nearly stone deaf, more than a little senile, and as eccentric as ever.

The rest of 1981 was relatively uneventful in terms of both camp life and the elephants. I got back from the United States in early March and went down to Amboseli soon after. I saw the T families many times over the next months. On March 23 I noted that Tara had developed small breasts and thus I knew that she was pregnant. On June 20 the TAs and TBs came and fed around the camp for about an hour and little Tuo, who was already learning bad tricks from his mother, came under the *makuti* (thatching) roof over my tent and drank the water out of my washbowl. Tuo was becoming a specialist at feeding just under the roofs of the tents, where greener grass grew because of the drip-off. The older elephants could not fit under the shelters but Tuo could, and since he had absolutely no fear, that became his niche in the camp. (A few years later he was still trying to get under the roofs to feed but he was much too big and he tended to have a devastating effect on the structures.)

On August 20 I found Teresia in estrus, which surprised me. I thought there would be more of a gap between the death of her calf and the resumption of estrus in a female of her age, or I thought she might not resume cycling at all. On that day M7, a one-tusked male, whom we named after Masaku, tried to mount her, but was unsuccessful. The next day Teresia was in consort with another male, Pablo.

Throughout 1981 I watched the gradual independence of Teddy. Some days he was with his family, other days he was with one or two young males, and on yet other days he was at the periphery of different families. By midyear he was spending less than half his time with his own family and I considered him independent. Teddy was 15 in 1981, which was actually a bit late for a male to leave. The average age of full independence is 14. Some males leave as early as nine years old, while others are real Mama's boys and stay until 19. Igor, in the IA family, was still closely following his mother, Isabel, when he was 18 and only became independent because she died. Albert finally left the AAs and his mother, Amy, when he was 19. There seem to be many factors involved in the timing of a young male's independence. The behavior of the females in a male's family may influence when he leaves. Some females are more tolerant of young males than others. Whether a male's mother is alive or not may also play a part in his departure. Then, too, a male's own character is probably an important factor; some males are less dependent and more adventurous from the time they are young calves. Gomer, of the GB family, was only nine years old when he began to go independent and yet his mother, Gloria, was alive and was the matriarch of the family.

Later in 1981 I began to see Right Fang away from his family and I sometimes found the TCs without him. He was 11 at the time, which seems to be the average age to start going independent. On the other hand, Tim, who was a year older than Right Fang and whose mother, Trista, had died several years before, showed no signs of leaving his family unit.

Thus, by the end of 1981, the only change in the number and composition of the four T families was the departure of Teddy. I did not consider Right Fang independent yet, but thought he would probably leave sometime in the next year or two.

Chapter VIII

Flexibility

1982

The TC and TD families had migrated away from the swamps during the wet season of 1982 but had not gone far. By June they were back in their dry-season home range. In July and August, along with much of the rest of the population, they spent a great deal of time to the south of the park in the *Balanites* and *Acacia nubica* woodland, feeding on the remaining grass and small trees. When that area was no longer rewarding they returned to the swamps and woodlands within the park, but there Teresia and Slit Ear found more elephants than usual and less food to go around. The "long rains" of 1982 had proved to be lower than average and the number of central elephants had grown through both births and immigration. There were more elephants concentrated in and around the two major swamps—Longinye and Enkongo Narok—than there had been in many years, perhaps ever before.

The TAs and TBs stuck close to the lodges and supplemented their diet with food from the rubbish pits. But Teresia and Slit

Ear used a different strategy to avoid the competition within the park.

In the early hours of one morning in late August Teresia decided to leave the park and seek forage elsewhere. The two families had spent the night in Olodo Are, to the east of Longinye, but instead of waking up and moving slowly toward the swamp, the group members stood around in some disarray. Teresia faced east, rumbling gently. Some of the others were facing west but they turned and went over to Teresia when she called to them. Soon all the family members were facing east and with a decisive slide-flap of her ears, Teresia started off. The two families walked east for about three miles without stopping to feed. Just before dawn they crossed the park boundary and passed a series of Maasai wells used for watering cattle. It was too early for the Maasai to be at the wells; they were barely stirring in their nearby thorn enclosures. The group walked on and eventually hit a dry water course which was the tip end of the westerly drainage of the Namalog swamp. Teresia, who was nearing 60 years old, had been coming to this swamp in the dry season from the time she was a young calf. It had never been the center of her family's home range, but her grandmother and her mother had led her there for four or five days of feeding at least once during each dry season. The vegetation was similar to that in the swamps within Amboseli Park but there were differences, including some tasty bushes and grasses which did not grow in either Longinye or Enkongo Narok.

When Teresia had come with her grandmother and mother, and even in recent years as matriarch herself, Namalog was just another swamp in the series of swamps at the northern base of Kilimanjaro. There had been no more Maasai at Namalog than anywhere else, although it was a particular favorite of the Maasai because the water at its source was considered fresh, clear, and unsalty. *Namalog* means "sweet place" in Maa.

On this day in August 1982 Teresia and the others walked along the dry course of the swamp among large *Acacia xantho-phloea* trees until they reached the main area of wet swamp, which was deep and thick with papyrus and other reeds. They skirted the deep part, heading along the northern edge for an area where the grass was usually good. Teresia deliberately kept to the northern side of the swamp because a few years before she had led her and Slit Ear's families to the springs at the south end of the swamp only to find fences and channels and smells of vegetation with which she was unfamiliar. She would have investigated further but there had been the hated sounds and smells of humans, cattle, sheep, goats, donkeys, and dogs. The unusual mix of odors had disturbed the elephants, and they had retreated.

Since that day, on all of their forays to Namalog they kept to the northern side or the western end of the drainage. In these areas they found only Maasai herding their cattle, and with discretion they were able to avoid direct confrontations. So on this day Teresia confidently led her group along the edge of the swamp in the safety of the thick groves of acacia trees. They had gone about halfway along the main deep swamp when Teresia suddenly stopped and lifted her trunk to smell the breeze. The wind was coming from the east so they were walking into it. Up ahead Teresia could smell people and domestic animals and a very unpleasant odor of vegetation that she had never smelled before. She gave a sharp shake of her head in a gesture of elephantine displeasure. The smells were coming from the area she had planned to feed in.

Teresia turned around and started back along the swamp and then led her family into the deep water and across to an island in the middle. Here they were able to feed without being disturbed. Over the next days, through smells and sounds, they learned where the Maasai were and how they were moving with their cattle. By retreating into the park at times and feeding at the swamp edges only at night, and in the deep swamps and on the islands in the daytime, the TCs and TDs were able to avoid the Maasai and use Namalog almost as their own private swamp. It was definitely more dangerous than the park swamps but at the same time there was more food at Namalog. Breaking their traditional dry-season pattern, they stayed in the east for over two months.

I "lost" the TCs and TDs for almost five months during 1982. I had seen them on June 2, 1982, and then I did not see them again until October 25, 1982. I started to get very worried about them by the end of September because of all the families they were two of the most central and usually fairly predictable in their ranging and movements during the dry season. In fact, by the beginning of 1982, of all 48 families in the population, the TDs were the ones that I had sighted most often during the course of the study—over 300 times.

I had expected not to see them in August. That was the time I always referred to as the elephant's "French holidays." (In France government offices and industries are closed down for the month of August and everyone takes a holiday at the same time.) It took me several years to realize that there was a definite pattern of the elephants disappearing in August. They would come back from the wet-season migrations in June, stick around for most of June and July, and then vanish in August. It was puzzling why they left again because there was no rain in those months and they still needed the swamps for drinking. Eventually I discovered that they were spending their time to the south in the *Balanites* and *nubica* woodlands, feeding on these trees and other vegetation, and presumably coming into the park to drink at night. By mid-September the majority of the central elephants were usually back in the park concentrating around Longinye and Enkongo Narok, where they stayed until the short rains came in November.

When Slit Ear and Teresia did not come back in September, I could not imagine where they were. I did not think they were all dead because usually only one or two elephants in a family die or are killed and then one finds the remnant of the group. In this case I had not seen any of them. During most of the months they were "missing" I was the only one on the elephant project so I thought it was possible they were around but that I was not running into them. I began going out and deliberately searching for them, but had no luck.

Keith had finished his master's degree at the University of British Columbia and had arrived in Kenya in June to start a new study on the feeding behavior of the Amboseli elephants for a Ph.D. at Cambridge University. Unfortunately, because of a particularly bad bout of bureaucratic red tape he was not able to get his research clearance until September. When he finally received permission to begin, I asked him to keep his eye out for the TCs and TDs. Keith saw them briefly on October 13. They were way over at the eastern border of the park. Keith had been to Namalog to ask the Maasai if the elephants were still using that swamp and they told him that one group had been there for several weeks. Circumstantial evidence suggested that it was Slit Ear and Teresia's families.

Keith and I were concerned about the situation at Namalog. The Maasai had begun to plant crops using the water from the swamp for irrigation. While they were pastoralists the Maasai were tolerant of wildlife, but we knew that as soon as they became agriculturalists they would come into direct and acute conflict with the elephants and other herbivores. At that time the Maasai were mainly planting onions, which so far the elephants did not seem to like, but they were planning to plant maize, cabbages, and beans and it was only a matter of time before the wildlife discovered these delicious crops.

I finally saw the TCs and TDs on October 25, and I believe this was the day they actually came back to central Amboseli. I found the two families coming in from the east to the northern edge of Longinye. Everyone was there except Tia, Tio, and Right Fang. Tad, Tia's 1976 calf, was with the group and I found that worrisome, because if Tia had temporarily gone off on her own Tad should have been with her. I was pleased, however, to see there were two new additions. Tara and Theodora each had new calves, both males. When I had seen the families on June 6, I had noted that Tara and Theo had three-quarter-sized breasts. I had predicted that they would give birth within the next couple of months. From the calves' sizes and levels of coordination, I estimated that Theo's had been born in July and Tara's in August. Earlier in the year, on January 27, I had recorded another new calf in the TC family. Tess had given birth to a female and I was glad to see that this calf was still alive and appeared healthy. Tess, at 19 years old, had her first surviving calf.

I was concerned about Tia, but the whole family seemed relaxed as they strolled in slowly and placidly. The only indication that they might have had some trouble while they were away was a new slit cut in Slit Ear's lower left ear. I drew it:

The slit, which began in the inner part of the ear, was unmistakably the size and shape of a Maasai spear. It must have hit her ear as she stood facing or charging a Maasai. There was no sign of a spear wound on her neck or shoulder behind the ear so she must have had her ears spread. I thought that Tia might not have been as lucky. However, the very next day I found the TCs and TDs and there was Tia with Tio and Right Fang. Thus, it seemed as if Teresia and Slit Ear's strategy had been a successful one. All the family members were alive and looked in fair condition, and the three calves born to the two families that year were doing well.

I must admit that it surprised me that the TCs and TDs would change their dry-season ranging pattern so drastically. I had come to expect a certain regularity in elephant behavior and I would never have predicted that these two families would leave the central Amboseli swamps for most of the dry season. But 1982 turned out to be a year when the whole question of elephant flexibility, both social and ecological, became of particular interest to me. I was doing much less field work and instead was sitting down and analyzing some of my data. One of the things to emerge was the intriguing ability of elephants to change their behavior under various ecological conditions. The second thing to draw my attention was an overall change in the migratory patterns of the elephants in the previous five years. This change, which was also being investigated by Jonah and Keith, posed some ecological and park-management questions.

Even before I began to analyze my data I realized how easy it was to come to some perfectly reasonable, but inaccurate, conclusions after watching a population of elephants for one or two years. It was very tempting to superimpose a firm structure to one's observations of social organization and ranging patterns and fit it all into a neat package. As I explained in the earlier chapters, the first part of my study covered a period of drought. At the time I was not aware of how serious the drought was or what a profound effect it was having on the elephants. During the part-time study from 1972–75 I was getting to know the elephants individually and working out who belonged to what family, what their ranges were, and how they were using the habitat. By the end of 1976, when I had been living in the park for over a year, I felt fairly confident that I understood the basic behavior and ecology of the Amboseli elephants. I was wrong. I did understand some of it, but after a year of good rainfall in 1977, I realized that I had only been observing the elephants under one set of conditions and that their behavior could change quite radically under another set.

In looking at my data on elephant social organization in 1982 I saw that creating a diagram such as the one I presented in Chapter V, however helpful, did not really reflect the immense flexibility within the elephant social system. The degree of stability of family units is one of the features of the system that reveals this inherent flexibility. The subject has even been the focus of a friendly controversy. When Iain Douglas-Hamilton did his study in Manyara he found that family units were remarkably stable in composition. When he came across Florence's family, all ten members would *always* be there. It was the same with Rhea's family or Leonora's or any of the other well-known families in the population. In his four-and-a-half-year study he saw one young female separated from her family for nearly a week and a gradual breaking off of the subgroup of five that I mentioned earlier, but these were the only exceptions in the daily records of family-unit composition. Sometimes two or more families would join together and coordinate their activities. These were usually members of a single bond group. Occasionally loose aggregations of up to 100 elephants would form and move through an area feeding. But when the aggregation broke down, the same animals would always remain in the same families.

The results of an extensive radio-tracking study carried out in Zimbabwe by Rowan Martin called into question Iain's description of the degree of family-unit stability. Martin found that a female was rarely in a group of the same size or composition and that the most consistently stable association was between a female and her youngest offspring. At the same time that Martin was carrying out his study in Zimbabwe, I was getting similar results in Amboseli. Family-unit composition in Amboseli was far less predictable than the way I had remembered it to be in Manyara.

Fortunately, it turned out to be a case of no one being "wrong" but of elephants behaving differently under differing ecological conditions. In the early days of wildlife research many of the *scientists* were inflexible. If someone studied a species in one place and someone else got incompatible results in another place, it caused all sorts of anger and backbiting. It was almost as though each researcher thought his or her study animals displayed the Platonic ideal of the species' social behavior and thus anything that contradicted his or her description *had* to be wrong. Primates in particular seemed to generate controversy. (Elephants generate intense controversy as well but not in the behavioral field as much as in the ecological: The mere mention of elephants and trees is usually enough to start a raging battle among wildlife biologists.) Gradually, as more and more results came in from a variety of places and habitats, scientists were forced to give up their entrenched positions. They soon became fascinated with the differences among populations and what they meant in terms of adaptive significance. Now the topic is very fashionable.

My analysis of family-unit composition and groupings over time in Amboseli showed that there was a great deal of variation among families and also within a family under differing circumstances. First of all, membership in a family unit did not necessarily mean that an individual would always be present in the family or that the family would always have the same composition every time it was sighted. Stability of composition varied from family to family with some tending toward fission within the family and others tending toward fusion with other families.

Taking the TC family as an example once again, my analysis indicated that Slit Ear's family was very stable in that all members were together in the great majority of sightings, but even this family was found in different *group sizes* on nearly a daily basis. As I said earlier, the TCs were with the TDs in over 80 percent of my sightings of them. In the two-year period 1980–82, before the births of new calves, they were often in a group of 16 animals, made up of 10 TCs and 6 TDs, but there were sometimes 1 or 2 young males with them and sometimes older males as well. On other days they might have joined Tania and Tuskless, adding eight more animals to the group plus whatever young males may have been tagging along with the TAs and TBs. At the same time Teddy was nearing independence and was sometimes present, sometimes not. Males of Teddy's age tend to come and go during the year or so before leaving their family for good, and their presence or absence changes the size and composition of the group from sighting to sighting. Then, too, the TCs occasionally joined other families in feeding aggregations.

Thus, one day Slit Ear might have been in a group of ten consisting of just her family, the next sighting she might have been in a group of 16 consisting of her family plus Teresia's, the next in a group of 26 including the TAs and TBs and two young males, the next she may have been in a group of six, having split with Tia for a brief time, and then very next sighting she may have been in an aggregation of 130. If one had been radio-tracking her from the air or stations on top of hills, these kinds of data would have presented a confusing picture. On the ground with Slit Ear and knowing all the other individuals (which they did not know in Zimbabwe), I could see what was happening. Slit Ear's family unit itself was stable, her splits with Tia were very rare, and when she aggregated with other families her own family was almost invariably intact.

Some other families in Amboseli were not as stable as the TCs and it was interesting to try to figure out why they might not have been. I looked at the records of the 24 best-known families in the Amboseli population. Eighteen of these families were fairly stable, with all the members found together in over 70 percent of their sightings. The six remaining families, whose members were found all together in less than 70 percent of their sightings, appeared to be unstable for a variety of reasons. I will briefly describe each of these less stable families in descending order of cohesiveness to try to illustrate some of the probable causes.

In 1982 the FAs' matriarch, Filippa, died, leaving two adult females, two juvenile females, and three calves. For nearly a year afterward the family was split in three. The two adult females, Fiona and Fifi, formed separate mother/offspring units and the juvenile females sometimes moved on their own, sometimes with one or the other of the two females, and sometimes with other families. Prior to Filippa's death the FA family was one of the most stable. In this case the loss of the matriarch appeared to have a disintegrating effect on the family. Without her leadership and the strong bonds the other females had with her, the family split. After a year Fiona and Fifi began to spend more time together again but the family still had more of a tendency to divide than it did before.

As I described in Chapter V, the largest family unit in the population, the PA family, with 29 members in 1982, gradually split into two units of 16 and 13. By the end of 1982 they were considered separate families, with the two largest and oldest females, Penelope and Phoebe, each leading one subunit. The size of the family unit appeared to be the cause of the split in this situation. I would guess that beyond a certain number the loss of feeding efficiency outweighs the social benefits of living in a large family.

The LA family lost an unnamed adult female in 1975. Two young females, Louise and Libby, who were the putative offspring of this female, often split from the other members and moved as a separate group. I assumed that their relationship with the matriarch, Lillian, was not a close one, but I was always puzzled by the split because not only did Louise and Libby go off but so did an adolescent female, Lee, who I was fairly certain was the daughter of Lillian. Perhaps as a small calf Lee had been taken care of by Louise, the older offspring of the dead female, and had built up a stronger bond with her than with her own mother. The bonds between the two younger animals did not surprise me, but a daughter leaving her mother did.

The KA family's matriarch and another adult female died in 1975, and there were members in the unit whose relationship was uncertain. Some may have been offspring of the two dead females, but others may have been immigrants from families who had lost all their adult females. I knew of one family whose two adult females were killed on the same day, leaving no one older than ten years. It was in the early days of the study and I had not yet photographed the youngsters, and therefore did not know what happened to them. I think in a situation such as this they would have tried to join a family within their bond group. In any case, "floaters," as I came to call these presumed orphans, never seemed to establish quite as strong bonds as they would within their own families and they tended to join and leave their adopted family for unpredictable periods.

The FB family also contained individuals who were floaters, and bonds among the members in general did not seem as strong as in the other families. One reason for this may have been the character of the matriarch, Freda, who is irritable, possibly because of a physical ailment—she has constant open sores on one side of her body. She often goes off on her own with just her youngest calf.

The DB family has always refused to fit into any neat definition of a family unit. I have often wished I could just ignore them when doing analyses. The DBs may not even be one family and they certainly have never been cohesive. In 1975 the "family" consisted of seven animals: two large adult females, two young females, one immature female, one first-year calf, and a pubertal male. In the next three years the two young females and the calf died and the pubertal male became independent, leaving only three members of the family: Delia, Deborah, and Daisy. Even then the levels of association of these three individuals were not high. Then, on November 15, 1978, Delia, one of the large adult females, had the first calf born to the Amboseli population after a 16-month period of no births. From that day the QB family, a peripheral family which previously had rarely associated with the DBs, began to spend time with Delia and the calf. At first Delia and the other large DB female, Deborah, were intolerant of the QB females, and often threatened them and chased them. The QB females, however, initiated greeting ceremonies whenever they approached the DB females, until the DBs began to respond by greeting in return. Eventually a close relationship built up between Delia and the matriarch of the QBs, Quilla. Deborah and Daisy remained more aloof.

By the end of 1982, Quilla and her offspring were spending most of their time with Delia and her offspring, and the rest of the QB family had resumed their previous distribution and association patterns. Deborah and Daisy were more frequently on their own or with the DAs, members of the same bond group, than with Delia and Quilla, and I considered that the DB family had broken into two. As I write this in 1986 I am still not sure what has happened. Quilla continues to spend a great deal of time with Delia but goes off on her own or joins up with the other QBs from time to time. Because there is evidence indicating that the central elephants have better forage available to them in the dry season, it makes sense on one level for Quilla to try to insinuate herself into a central family. However, where does that leave all the advantages of living in a group of related females in a system that evolved over thousands of years? Whether Quilla will become a permanent member of Delia's unit is yet to be seen. If she does it will be the first recording in Amboseli of a female leaving her own family and permanently joining another.

In each of these six cases there seems to be a different combination of factors accounting for the lack of family stability: the death of a matriarch, the size of the family, the strength of the bonds among the members, the presence of floaters in the family, the physical condition of an important member, and the attempted immigration of a stranger. And there are probably many more things of which I am unaware.

So far I have pointed out possible social reasons for lack of stability in some family units, but perhaps more important and more prevalent are ecological reasons. As the long-term study progressed, it became obvious that the cohesiveness of families was greatly affected by environmental conditions. Once again I appreciated the value of several years of data. I was able to compare association patterns within families and among families under varying conditions. I chose 1976 and 1979 as contrasting years, the former a year of severe drought and the latter of higher-than-average rainfall.

The most obvious contrast for anyone even superficially watching Amboseli's elephants in those two years was the difference in the size of cow/calf groups that one encountered. A "group" was defined as any number of elephants moving together in a coordinated manner, ranging from one (extremely rare) to hundreds, consisting of females and calves or females, calves, and adult bulls. (All-bull groups were analyzed separately.) In 1976 group size was small, averaging only 15.1, while in 1979 it was 45.9. The distribution of group sizes in those two years was even more revealing. In 1976, 86 percent of the groups sighted consisted of fewer than 25 animals and only one group of over 100 was seen. In 1979 only 60 percent of the groups numbered under 25, and 16 percent consisted of large aggregations of over 100 elephants, many of which were over 200 and 300.

In Amboseli what appears to happen is that elephants will get into big groups when environmental conditions permit. In 1979 seventeen inches of rain fell following two years of high rainfall (average is about 12 to 13 inches) and as a result there was abundant, widespread, and evenly distributed vegetation. Under conditions such as these elephants can move and feed together, all getting basically the same food with little or no competition among them. One day I recorded a group of 550 elephants: It consisted of 47 family units and well over 50 independent bulls.

In 1976 only seven inches of rain fell after several years of average to lower-than-average rainfall. As a result, food resources were scarce. Since elephants had to search for and pick and choose their food from among patches of grass, bushes, trees, and swamp vegetation, it appeared to be more efficient to forage in small groups. The elephants spread themselves throughout the park and surrounding areas, with each family returning to its clan's dry-season home range and tending to move on its own. During the severe drought months of 1976 even some of the family units broke down, with a single female and one or two of her offspring forming a subunit. It appeared that younger females or females less closely bonded to the matriarch were often the ones who split, such as the LA family's little subgroup.

In Chapter II, I already described one set of circumstances that could cause fission—the birth of a calf. However, the ecological conditions prevailing at the time seem to be an important factor in determining whether a new mother stays in her family or not. I looked at the differences in this behavior between 1976 and 1979. Out of 21 females who gave birth in the 24 well-known families in 1976, 12 (57 percent) left their families for periods ranging from a few days to two months. In contrast, in 1979 very few females split from their families. Only five new mothers out of 57 (9 percent) separated from their families that year and four of these for only a day or less.

Fusions of families also occur. Temporary unions take place frequently, depending on environmental conditions. During the wet years, 1977–81, some of the bond groups remained in large extended families for long periods. The TAs and TBs were nearly always together as were the TCs and TDs. If I had first recorded the TCs and TDs in 1977 I would probably have considered them one family, but in the early years I had seen them as separate units often enough to decide they were two distinct families.

Permanent fusions may also happen. From the beginning of the study, the BB family made up a bond group with the UA family, but their association levels were not particularly high until 1977. In 1976 they spent only about 25 percent of their time together, but they still spent more time with each other than with any other families in the population. In 1975 four of the UA family members (including the matriarch) died, leaving only three members. Once the drought ended the UAs' three remaining females gradually began to spend more and more time with the larger BB family. By 1979 the BB and UA families associated with each other over 80 percent of the time. This level was as high as some of the individual female-female associations *within* families, and at that time I considered that they had fused.

After a virulent pneumonia-like disease swept through the Lake Manyara elephant population killing over 100 animals, Rick Weyerhaeuser, working under Iain's direction, recorded two cases of fusions of remnant families. He also noted that during and after the 1976 drought and the period of disease-related mortalities, some family units broke down, splitting into as many as three subunits. Thus both fission and fusion were recorded in the one other elephant population that was well known.

I think it is fairly easy to understand the environmental pressures that cause aggregations to break down and even family units to split up from time to time, but what I am not confident about is why elephants get into large groups at all. Just because they can do it is not a good enough explanation. It has been suggested that elephants move in large herds when they have been disturbed by poaching or other harassment. One theory is that when an elephant population is heavily poached, the old matriarchs are killed because they have the biggest tusks, with the result that the leaderless family units get together and mill about, not knowing what to do. A simpler theory is that an individual is less likely to be killed in a big group by mathematical probability alone, and thus those elephants that tended to mass together in times of danger survived and passed on this herding behavior. I have been to areas where elephants were under intense poaching pressure; on one recent occasion I saw a group of over 300 elephants at the height of the dry season. Although they were immediately undisturbed and were moving slowly, they were bunched very tightly together, forming a huge gray wall of elephant hide. Poaching was reported to be rampant in that park, and I had no doubt that this large aggregation had formed in response to predation.

In Amboseli, herds of over 60 are rare in the dry season. And when the big aggregations form during the wet season, they are usually loose and somewhat formless, with spaces between individuals and subgroups. In the years when I saw the biggest groups in Amboseli there was no poaching and a minimum of spearing and thus the families were not losing their matriarchs or being harassed. In Amboseli there had to be another reason for the formation of large aggregations.

I can only suggest that there are probably many factors that attract elephants to one another. Elephants are very social animals. The 650 elephants in Amboseli all know one another; in their fluid social system, where they all move about the ecosystem, sometimes together, sometimes apart, relationships of some sort must be established between individuals and families. Thus the big aggregations may provide an opportunity for reestablishing social relationships, whether they be affiliative bonds or dominance positions. I noted a high rate of interaction among the members of the aggregations especially in the first year that they formed. There were frequent greetings, friendly investigations, play bouts, threats, confrontations, pokes in the backside, and chases. Young males who had not yet left their families were the most active, taking every opportunity to meet other male calves in order to spar and test each other's strength. This sparring as youngsters is no doubt important in determining dominance rank in adulthood.

Large herds may also form for breeding purposes. The aggregations attract adult males, particularly males in musth. At the same time the social stimulation of a big group may bring females into estrus. In a large group an estrous female would have many more males to choose from and she would be more likely to find a musth male to go into consort with. In 1977 and 1978 large aggregations were frequent and during those two years 60 percent of the females conceived.

And finally, elephants are intelligent, complex animals who may simply enjoy the social stimulation of a great gathering of relatives, friends, acquaintances, and members of the opposite sex. I think this stimulation may be the proximate cause for getting together, with the ultimate causes relating to individual reproductive success.

Their social flexibility makes elephants particularly fascinating to study. After 14 years with the Amboseli elephants I still feel a sense of excitement when I head out of camp in the morning. I can never predict what I am going to see, and in all my years in Amboseli, I have never once been bored when I have been out with the elephants. One of them will always do something or be somewhere or be with someone that perplexes me. The combinations, recombinations, offshoots, splinters, breakups, and just the general mix-up of individuals are endlessly intriguing.

At about the same time that I was looking at my data on family-unit cohesiveness and flexibility, Jonah and Keith were analyzing grouping and ranging patterns of the Amboseli elephants using data collected from Jonah's aerial counts and Keith's, Jonah's, and Harvey's radio-tracking records. They eventually wrote a joint paper called "Seasonal herd dynamics of a savanna elephant population" for the *African Journal of Ecology*.

They too found that group size changed according to the season, with large groups forming during the wet months and small groups during the dry months. They also showed that during the rains the elephants ranged much farther than in the dry months. The elephants left the Amboseli basin shortly after the rains began and returned when water supplies ran out in the bushlands. In the dry season the radio-collared animals were always found within ten miles of the Amboseli basin while during the rains they might be up to 25 miles away. The elephants tended to use the bushed grassland outside the park when there was available water and abundant forage. As the dry season progressed, they moved successively to the open woodlands, the denser woodlands, the swamp edges, and finally into the swamps themselves. The quality of the forage in these habitats went from high in the bushed grassland down to the lowest level in the swamps. At the end of the dry season the elephants had to trade quality in the form of high-protein grasses, herbs, and shrubs, which are seasonal, for quantity in the form of low-quality but abundant swamp vegetation.

This pattern of leaving the basin during the rains and returning through the various habitats in the dry season was fairly consistent up until 1977. Then, being elephants and therefore not predictable, they changed their seasonal migrations. Once the good rains of 1977 came, the elephants tended to stay in the basin more and more. In the wet years of 1977–81 they were found in the basin in both wet and dry months, whereas in the dry years of 1973–76 the majority of the elephants migrated out in the wet season. The tendency to remain in the area of the park meant that there was a greater impact on the vegetation year-round, and it was suggested that this impact could lead to a so-called elephant problem, in which people perceive a conflict between elephants and their habitats. For the time being, the big question was why they changed their behavior.

Jonah, Keith, and I have suggested a variety of answers to the question, but none of us can be sure of any of them. The most obvious change that occurred in Amboseli in 1977 was the removal of the Maasai and their livestock from the 150 square miles designated as national park. As I explained in Chapter II alternate water sources plus other facilities were provided for the Maasai in exchange for their giving up the central basin. The other notable change was the substantially improved environmental condition of the park after the breaking of the drought in early 1977. There was plenty of grassy forage available for the elephants and they may have decided it was not worth the effort to move away from the basin. At the same time their decision may have been influenced by the poaching that had occurred on their wet-season migrations in the mid-seventies. Possibly they avoided going to the areas where elephants had been shot, once they had the choice. In the dry years they had to migrate in order to find nutritious food, but in the wet years there was sufficient forage "at home," with the added benefit of no Maasai livestock competing for it.

I personally feel that the change in the migratory patterns of the elephants was influenced less by the availability of vegetation than by the new distribution of the Maasai people. Having spent thousands of hours with the Amboseli elephants, I know that what they fear most is the Maasai. The elephants will panic and run when they see or smell them at a distance of over half a mile. Before 1977 Maasai were everywhere throughout the elephants' range; but after the park was created there was an area where the chances of meeting them were relatively lower. I say "relatively lower" because the Maasai still came into the park under a variety of circumstances (when the water pipeline or boreholes were not working; in dry years when there was little or no forage left outside), although they were not allowed to set up their *enkangs* (thorn enclosures with houses inside) in the park.

But as I keep saying, once one decides that elephants are doing something for one reason or another they do something totally unexpected and counter to one's theory. By going to Namalog, where there were many Maasai in the dry season of 1982, Slit Ear and Teresia acted contrary to my theories, and yet what they did may actually have been a clever strategy since the benefits of fewer elephant competitors appeared to outweigh the costs of meeting Maasai.

Once they got back to Amboseli after the short rains had begun in the third week of October, they did not stay in their normal dry-season home range for long. In November, for several days in a row, they came from the ridge above the basin into the northwestern end of the Enkongo Narok swamp, an area I had seen them in only one or two times before. During that month Tia, along with Tio, Tad, and Right Fang, broke off from Slit Ear and stayed around the Longinye swamp. This fission in the TC family was also unusual, but Tia must have had her reasons.

During 1982 I noted some other events and milestones in the Ts' lives. Right Fang continued to show signs of going independent and Tim was finally seen away from his family for the first time in the beginning of that year and on several occasions again before the TCs and TDs "disappeared." The day I found them coming back into Amboseli Tim was not with them and I did not see him with his family on the remaining sightings in 1982. It appeared that Tim had become independent very quickly once he left the family for the first time. This behavior was in contrast to that of both Teddy and Right Fang, who came and went for a long time before finally becoming independent. In fact Right Fang was still with his family more often than not in 1982, so by the end of that year I did not yet consider him independent.

During 1982 I was watching the calves who had been born in 1980 to see when their tusks would begin to show beyond the lip. It was part of the study of calf development. To know the age of tusk eruption was helpful in formulating an aging technique for scientists working on elephant populations in which birth dates were not known. In Amboseli, somewhere around two and a quarter years old seems to be about the average age for tusks to emerge. In males, the tips appear earlier.

On March 31, 1982, I could see the tips of Tuo's tusks showing. He was 21 months old. I was particularly interested in this record because I had wondered if Tuo would inherit tusklessness from his mother. In Asia, probably through selective hunting for ivory, female elephants only rarely have tusks, and males sometimes have tusks and sometimes do not. In Africa most elephants have tusks: Tusklessness is extremely rare in males and not common in females. In Amboseli less than 2 percent of the population is tuskless and all are females; about 3 percent are one-tusked and about half of these are male. Tusklessness usually runs in families, such as in Big Tuskless's family, with (in 1982) three tuskless females and two one-tuskers; and in Gloria's family, with two tuskless females and two one-tuskers.

Estella's twins provided a convenient example of the difference in tusk development between male and female calves. I noted that Equinox, the male twin, had tusks on October 26. A month later his sister Eclipse showed just the very tips of her tusks, while Equinox's tusks already protruded at least a quarter of an inch beyond the lip. I estimated that Equinox's tusks must have erupted at about two years and two or three months old, whereas his sister's did not erupt until she was two years and five months. They were both a bit late compared to other calves of their respective sexes, but then both twins had all along been small for their age.

On November 19 I found the TAs and TBs in Longinye, together with a few independent young males including Tim. Tonie, the female who had stood over and protected her dead baby from lions two years before, was there with a newborn, healthy male calf. I was pleased for her and quite impressed with how quickly she had bred again. The first birth occurred in November 1980. Joyce had recorded Tonie in estrus on February 10, 1981, only three months later, and this must have been when she had conceived her new calf.

There remained only two young adult females in the T families who had not yet had calves: Tilly and Tamar. I was watching their breast development. During 1982 they had both gone from being completely flat-chested to having small rounded breasts about one quarter the size of a lactating female's. Neither Tilly nor Tamar had been seen in estrus, but it is easy to miss an estrous period that lasts only four to six days. Judging by the size of their breasts, I suspected they would have their calves sometimes in mid-1983, which meant that they had conceived in the latter half of 1981.

I was eager to gather data on Tilly and Tamar's reproductive activity because they were nearly positively known-aged animals. I had met them when they were both under three years old in 1973 and I was able to place their ages confidently to within a year. I estimated Tilly to have been born in 1970 and Tamar in 1971. Thus, in 1981, when I suspected that they had conceived, Tilly was 11 years old and Tamar was 10. I wanted to know at what age females reached sexual maturity and started to cycle, at what age they first conceived, and at what age they first gave birth. If my calculations were correct, Tilly would be 13 when she had her first calf and Tamar would be only 12.

The last months of 1982 were busy and exciting ones for the elephant project, the camp, and me. As I mentioned earlier, Keith returned in June to start the research for his Ph.D. on feeding behavior of elephants. Phyllis came back in July to begin work on maternal behavior and calf development. Keith's research clearance came through in September and Phyllis's in October and they were both raring to go. Keith already knew many of the animals from his previous study but Phyllis had to start from scratch. They spent those first months becoming acquainted with the elephants, taking new photographs, updating the recognition files, and working out their sampling techniques. They were very keen to contribute to the long-term records both in providing data and in devising filing systems. In this way and many others they were extremely helpful to me and I was very happy to have them on the project.

I had been working on my own for more than a year after Joyce had left in July 1981. I missed her companionship, of course, but what I missed most was someone to talk to about the elephants, someone who knew them personally as I did. Every evening when Joyce and I had returned from our field work we would immediately discuss which elephants we had seen, where they were, whom they were with, and what they were doing. Now I had two other people out in the field and each evening when we all got back it was like a great gossip session. "You'll never believe where I saw Jezebel, and guess who she was with!" "Lolita came into estrus today; she seems much too young, and all the dinky little bulls were chasing her nonstop." "M154 has just come into musth and he was testing all the females in the PA family." "I just missed seeing a birth today. Odile was with a tiny shaky baby and she still had blood on her legs." "Oh, did you get the sex?" "Yes, it's a female."

With Phyllis and Keith in the field I felt less reluctant about starting my new role in the project. The Kenya government had decided that there should not be more than two expatriate researchers working on a wildlife project at one time. I wanted both Keith and Phyllis to do their studies: The feeding study was a top priority and it was a perfect time for the continuation of the calf-development study since we were expecting another baby boom in 1983. I knew that what I had to do was write up my results to date, and so I decided to give up my research permission to make way for Keith and Phyllis. At the same time Joyce agreed to delay a study on communication that she wanted to do. She had completed her Ph.D. in October 1982, having set a record at the Sub-Department of Animal Behaviour at Cambridge for the shortest time taken to obtain a Ph.D. degree. But more important, her thesis was considered an excellent piece of work by all concerned.

My new role was to act as director or really coordinator of the project. I was concerned with keeping the project and the research camp going and collecting and organizing the long-term data on births, deaths, estrus, musth, group size, and ranging and association patterns. The Wildlife Department and the Office of the President, which grants research permission, agreed to this function and allowed me to continue to spend time in Amboseli to supervise and confer with my colleagues. It was a good solution, but a part of me was very sad to give up the more intensive research.

My plan was to buy a computer and spend additional time in Nairobi. Both Keith and Phyllis were living in the camp. The romance that had been budding back in 1979 when they had been in Amboseli had flowered, and before 1982 was over they decided to get married. (They had a delightful wedding in March 1983 in Tsavo National Park, at the house of our good friends Joe and Christina Kioko. Joe, who had been warden of Amboseli from 1974–78, had since been transferred to Tsavo.)

So 1982 ended on a good note for the project and for the elephants. Despite the dry year there had been very few deaths. Only ten elephants had died: eight calves and two adult females. The Ts had done very well. With the new births and the departure of Tim, the T families consisted of the following 26 animals:

TA	Tuskless (62)	♀
	Tuo (80)	♂
	Tonie (67)	♀
	C'82	♂
	Tilly (70)	♀

TB	Tania (44)	♀
	Tom (79)	♂
	Toby (73)	♂
TC	Slit Ear (36)	♀
	Slo (80)	♀
	Tabitha	♀
	Tamar (71)	♀
	Tara (67)	♀
	C'82	♂
	Tess (63)	♀
	C'82	♀
	Tia (50)	♀
	Tio (80)	♂
	Taddeus (76)	♂
	Right Fang (70)	♂
TD	Teresia (22)	♀
	Tolstoy (71)	♂
	Theodora (67)	♀
	C'82	♂
	Tallulah (63)	♀
	Tao (80)	♀

Chapter IX

Population Dynamics

1983

Three new babies had been born in the T families by June 1983. Tania gave birth in April and Slit Ear and Tia had calves in May, once again synchronized in their reproduction. These new calves did not cause quite the excitement that Tom, Tao, Tio, Slo, and Tuo had in 1979 and 1980. When the '83 calves arrived there were already nine calves under four years old in the bond group and only one adolescent female to shower the usual intense attention on the newborns. Slit Ear's eight-year-old daugher, Tabitha, spent as much of her time as possible close to Slit Ear's and Tia's calves, touching them frequently with her trunk, and pulling one or the other in under her chin to protect it. She always stood patiently whenever one of them wanted to suckle from her flat, milkless breasts.

Slo and Tio, the 1980 calves of Slit Ear and Tia, did not welcome the arrival of their siblings. Their mothers refused to suckle them any longer. Slo and Tio were only just over three years old, which was relatively young to give up milk, and the weaning process was difficult for them. Slo gave in to the

231

inevitable but Tio kept attempting to suckle. He would try by following close behind Tia and hooking his trunk around her hind leg in a calf's signal for its mother to stop. If she stopped he would then go forward to her teat and if she was cooperative she would stretch the leg on his side forward so he could easily reach her breast. If she did not stop or was uncooperative in standing, which she had been ever since the new calf was born, he would scream in protest. In the past she had always given in to him when he had a tantrum, but now she would not. Sometimes when the baby was suckling he would try to sneak onto the other teat, and once in a while Tia would tolerate him, but within a few weeks he gave up trying. Fortunately, it was the height of the rainy season and there was sweet new grass in abundance, providing Slo and Tio with an adequate diet. In the meantime, Tom, who was closer to four years old when Tania had her new calf, made the transition far more easily. He had always been such an independent, outgoing calf that he did not seem to mind the arrival of the baby.

Two of the young females in the Ts, who were usually very attentive to new calves, barely acknowledged the '83 babies. Tilly and Tamar were noticeably bulging, heavy and slow, and their breasts were now over three quarters the size and fullness of those of lactating females. Tamar, who was just 12 years old, appeared to be the one more encumbered by her pregnancy. She stopped and rested frequently and lagged behind her group.

On the evening of June 20, 1983, the TCs and TDs were feeding in the *consimilis* to the east of Longinye, getting ready to move out to the ridges beyond Olodo Are to feed at night while there were no Maasai at large. Tamar had been showing signs of restlessness and agitation. Like Tallulah, she stopped and backed and knelt down and also stretched her front legs forward and rocked back and forth. She appeared to be in considerable discomfort. She lay down, got up, knelt on her back legs, circled, and stretched over and over again, but no bulge appeared beneath her tail.

At 7:00 P.M., when the last light was fading in the western sky, the elephants formed a single group, and after a few rumbles were exchanged they set off to cross the open pan. Tamar did not follow them. The two families had walked about 300 yards when Slit Ear, with her C'83, stopped, turned back slightly, and rumbled. Tamar answered. Slit Ear rumbled again but Tamar remained quiet. The others were continuing on across the pan and Slit Ear began walking to catch up with them. She went another 100 yards and stopped again. Once more she rumbled and Tamar answered. Slit Ear turned around, facing the *consimilis*. She lifted and swung her right foot back and forth, and rumbled a louder sound. There was no answer from Tamar, but Tara and her C'82 plus Tabitha and Slo broke off from the others and turned and came back to Slit Ear. When they arrived where Slit Ear was still standing, they greeted her with flaps of their ears and the rich greeting rumble. They then all walked back to Tamar.

Tamar was lying on her side breathing hard. Slit Ear felt her with her trunk and then poked her hard with a tusk. Tamar struggled up onto her feet and immediately started backing and stretching and kneeling. Slit Ear, Tara, and the others, except the baby, began feeding on the new shoots of *consimilis* at the base of the large clumps. Slit Ear left Tamar alone as long as she did not lie down. Whenever she did, Slit Ear came over and forced her to stand up.

This behavior went on for over two hours and still no bulge appeared. Tamar began to lie down more frequently and when Slit Ear made her get up she stood with her head low and swayed unsteadily from side to side. She no longer moved about in an agitated manner. Finally, around 10:00 P.M., with a gush of blood a bulge appeared below Tamar's tail. Now she knelt and stretched and strained and at 10:15 the calf was expelled onto the ground. Only bits of the fetal sac still clung to it. It was covered in blood. Tamar stood swaying on shaky legs while Slit Ear and Tara tried to get the baby to its feet. They nudged it and attempted to lift it with the aid of their feet and trunks, but to no avail. The baby barely struggled at all and soon stopped moving. After a few more minutes it died.

Blood continued to flow steadily from Tamar's vulva, forming a pool between and around her hind feet. She began to quiver all over and then sank onto her knees and head. Slit Ear and Tara raced over to her to prop her up. They managed to get her up onto her feet again and with one on each side of her they kept her up for several more minutes, but once again she fell forward onto her forehead and tusks, and this time they could not get her to stand again. She fell sideways and lay quietly with just her trunk bending and turning. Slit Ear got behind Tamar and wedged her tusks under her shoulder, as Teresia had done with Tina, and lifted, but there was almost no response from Tamar, only a brief movement of her feet and trunk. Slit Ear tried to lift her three more times but then gave up and stood over her, with her trunk touching Tamar's head. Tara and Tabitha also stayed close to Tamar, touching her from time to time with a foot or trunk. Slit Ear's C'83 nestled in beside her mother and Slo waited at Slit Ear's tail, leaning against her but with her ears lifted and tense in confusion and fear.

It took Tamar a long time to die, but her mother and sisters did not leave her until well after she had stopped breathing. Very slowly they moved away from her, heading in the direction Teresia had taken, but they stopped often, turning their heads slightly back and rumbling. They got no answer from Tamar. After they had covered about three quarters of a mile, they halted once again and rumbled, and this time they got an answer from Teresia and Tia up on the ridge. Slit Ear, Tara, and Tabitha all raised their heads and rumbled a louder, more resonant sound and with a new eagerness to their gait they walked on to join the rest of their family and bond group.

Births and deaths are the main data points of interest to zoologists who are interested in population dynamics, which is the study of what a particular defined number of animals, or human beings, for that matter, is doing, has done, and will do in terms of rates of growth or decline or the reaching of some sort of stability. It is the cumulative effect of individual births and deaths, age of maturation, intervals between births, number of years of fertility, immigration, emigration, and so forth, on the size of populations. The topic often seems cold and dry, full of bare facts and figures when viewed in the abstract, but it is an essential one in understanding the ecology and behavior of a species.

Tamar's death and the birth and death of her calf were what my colleagues and I came to call "demographic events" when we were trying not to be emotionally affected by the death of an elephant we had known well. Whenever I arrived in Amboseli after a stint of writing in Nairobi, inevitably the first thing I wanted to know was whether there had been any births or deaths. "Have there been any demographic events?" I would ask lightly, while tensely waiting to hear the answer. I always felt happy when I heard about a birth and very sad when I was told that an elephant had died, especially if it was an individual I had known for many years.

Nevertheless, the demography of the Amboseli elephants is one of my major interests and one of the most important aspects of the long-term monitoring of the population. No other study of elephants in Africa has ever been able to record the data needed to describe the real dynamics of a population. Usually a scientist is forced to take a subsample of a population at one point in time and then extrapolate from that. But in Amboseli we know, first of all, the whole population, which is unusual in itself, and over a period of 11 years we have recorded virtually all births and deaths, age at sexual maturity, calving intervals, and the fecundity and mortality of animals at different ages.

In approaching a demographic study, one first has to try to define the population. I was fortunate in that Amboseli's elephants formed a fairly discrete population geographically. They did not always, but by the time I started the study in 1972 they were more or less isolated from any surrounding elephant populations. Historical records show that there used to be elephants to the west, concentrated around the mountain called Ol Donyo Orok, which is near the town of Namanga and 45 miles from the center of Amboseli. It is likely that these elephants met up with Amboseli's from time to time. Human expansion and poaching forced out Namanga's elephants and there are no longer any there. Some of them may have moved into Amboseli.

No elephants lived to the north except during the wet season, when Amboseli's own elephants migrated there and possibly interacted with some of the elephants from Tsavo National Park, which is to the east. Jonah's aerial surveys showed that Amboseli and Tsavo elephants overlapped slightly in their ranges but rarely at the same time, so there was probably little mixing of the two. In more recent times the Amboseli elephants have not been going far enough north or east to meet the Tsavo ones. Also to the east are the two swamps that I have mentioned before, Namalog and Kimana. These are within the Amboseli ecosystem and the elephants that use these swamps are part of the Amboseli population.

There is a population of elephants to the south of Amboseli living in the forests on the slopes of Kilimanjaro. A student from Tanzania's College of African Wildlife Management did a survey of these elephants and estimated them to number about 1,200. Ever since I have been working in Amboseli the Kilimanjaro elephants, or "Kili elephants," as they are known to us, have been coming to Amboseli on visits. They usually arrive in the wet season and stay for only a few days. I have speculated that they probably come for salt, which there is an abundance of in Amboseli and which is usually scarce on the slopes of mountains.

When they are in Amboseli the Kili elephants mix with the Amboseli ones, sometimes amicably and sometimes with exchanges of threats and other forms of aggression. Their appearance and behavior is totally different from that of the Amboseli elephants and therefore there is no mistaking them. The Kili elephants are, and I do not hesitate to say it, extremely funny-looking, verging on being ugly. They are smaller and leaner, have narrow pinched faces, with wrinkled foreheads and hairy heads, little triangular-shaped ears, skewed, oddly shaped tusks or no tusks, and squiggly, crooked tails which rarely have any hair at the end. In the first five years or so of the study they were extremely wary of vehicles on their visits and would panic and run away or charge any time a car came within 100 yards of them. In the last few years they have calmed down somewhat and a few of the Kili families and bulls are becoming relaxed about tourist and research vehicles. However, most of them still gather in a tight formation when they see my Land-Rover, lift their heads, and threaten me, with their beady little eyes glaring from over their misshapen tusks. I have grown rather fond of them because they always make me laugh.

Although the Kili elephants mix with the Amboseli elephants, there is every indication that they are two distinct populations. We have seen a few cases of Kili females being mated by Amboseli bulls, but this must not be a common occurrence since otherwise the general appearance of the elephants from the two populations would not remain so distinct. Phyllis and Keith found the carcass of a first-year Kili calf that had died in Amboseli and the lower jaw (which we collect for aging) was *much* smaller than the jaw of an Amboseli calf of the same age. On this evidence taxonomists might even suggest that the Kili elephants are a different race or subspecies of *Loxodonta africana*. It certainly looks as if they have been isolated from the savanna populations for long enough to have changed in size and body shape, possibly as an adaptation to living on forested mountain slopes.

Thus, except for infrequent contact with elephants from the east and south, Amboseli's elephants are self-contained and easily definable as a population. Jonah's aerial counts in the early seventies indicated that there were 600 to 700 elephants in the ecosystem. In 1972, when Harvey and I started the study, our method of finding out how many elephants there were in the population was to photograph and register every individual. Of course that took a few years. By 1974 we had registered the majority of the cows and calves and less than half of the bulls. I had also been recording births and deaths, but because of the part-time nature of the study I am sure I missed many.

Working backward from how many elephants there are in 1987, when every single individual is known, and taking into account all the births and deaths per year since 1974, I have arrived at a minimum figure of 602 elephants at the beginning of 1974. This is a minimum because there were many elephants, especially bulls, who were never registered.

Joyce and I examined photographs taken of the Amboseli elephants from 1964–66 by ecologist Norman Myers, who at that time was a professional photographer. Many of the females were ones that I knew as older animals in the seventies and eighties, but we could recognize only one of the numerous males that he photographed. All the rest were gone by the time we started taking pictures of the Amboseli bulls. There were at least 20 big, distinctive, large-tusked males in those photographs and we both felt disappointed that we had never seen and known them. Norman Myers's photographs gave us some indication of the rate that the bulls were lost to hunters and poachers in those years. I also examined photographs taken of the Amboseli elephants by Mark Boulton (a conservation-education specialist who is an excellent amateur photographer) in 1968–69 and could see that there must have been a baby boom at that time because there were many young calves. There were also some females in the pictures that I had never known. Taking all this into consideration I estimate that the Amboseli population at the end of the 1960s was at least 650.

When I began the study in 1972 the Amboseli population was in decline and continued to be for the next six years. Starting from January 1, 1974, when our records for birth and deaths were adequate, to December 31, 1978, there was a minimum net loss of 124 elephants. These are the figures:

December 31	Number	Loss or Gain
1973	602	
1974	575	−27
1975	551	−24
1976	502	−49
1977	479	−23
1978	478	−1
	Total	−124

From the beginning of 1979 the population began to increase, and it continued to grow for the next five years, with a net gain of 196 animals. Below are the figures for those five years:

December 31	Number	Loss or Gain
1979	525	+47
1980	570	+45
1981	581	+11
1982	604	+23
1983	674	+70
	Total	+196

All the gains during this five-year period were due to an excess of births over deaths and not to immigration. In the previous six years all the losses were due to an excess of deaths over births and not to emigration. There is no evidence to indicate that any elephants have come into the population from Tsavo or Kilimanjaro or that any have left to join these other populations. When an elephant disappears we have every reason to believe it is dead.

I have broken these figures into two sections to illustrate how easy it is to make false assumptions about what is happening to a population of such long-lived, slow-reproducing mammals. A five- or six-year period is short-term for an elephant in a life spanning over 60 years. Nevertheless, there is no doubt that the population was in serious decline during most of the seventies and if the rates of loss had continued at, say the level of 1976— 9 percent per year—the population would have become extinct sometime in the early twenty-first century.

On the other hand, the rates of growth during the early eighties might have been considered alarming to wildlife managers concerned with the possibility of "overpopulation." In 1983 alone the Amboseli elephant population grew by 11.6 percent. But this percentage and the 9 percent loss in 1976 are meaningless in themselves. Fluctuations from year to year can be great, as well as very misleading. Unfortunately, crucial decisions such as the initiation of culling programs, are often based on data collected at one point in time. The Amboseli project is unique in having long-term population data compiled over 14 years. In many ways it is probably the most important work that we have done because the Amboseli study shows that elephants are flexible in their reproductive behavior and responsive to the various conditions that they find themselves in.

We were not the first to show this reproductive flexibility. Richard Laws had worked on culled elephants in Uganda, Kenya, and Tanzania and found that there were significant differences in many of the reproductive parameters between populations, all of which could profoundly affect the population dynamics. For example, in the Budongo Forest Reserve in Uganda he discovered that the age of puberty for females was delayed until 22 years old and the average interval between the births of calves for a single female was close to eight years. The retarded age of puberty and the very long calving intervals meant that this population could not possibly replace deaths (natural or otherwise) with enough births and so it was in decline. At the opposite extreme were the elephants of Mkomazi Game Reserve in Tanzania. Here the females were reaching puberty at 12 years old and the average calving interval was three years old. This population was growing.

Laws correlated these differences to population density— that is, the number of elephants per square mile. In Budongo there were 7.5 elephants per square mile, while in Mkomazi there were only 2 per square mile. The relationship between reproductive parameters and density appeared to correspond in all the populations that he studied. However, in Manyara, where Iain Douglas-Hamilton worked, the correlation did not hold. There the elephants lived at a density of 14 per square mile and yet they were a healthy, increasing population, with females reaching puberty at 11 and with calving intervals of four years. The difference appeared to be that Manyara's habitat was exceptionally rich and could support a high density of elephants. Therefore, environmental conditions as well as density were a key to understanding the regulation of elephant populations.

There has been a good deal of controversy over the regulatory processes at work within animal populations. Simply put, one view holds that self-regulation occurs in populations according to density-dependent factors; the other view holds that populations fluctuate in response to external environmental conditions such as droughts, famines, and predation. In a self-regulating population there is supposed to be an ideal point of stability, but it is seldom reached in the real world. In fluctuating populations animals increase when they can and decline when conditions are unfavorable; there is no inherent stability. At one time the extremes of these two views were advocated, but more recently a sort of melding of the two has occurred, and most people seem to agree now that populations are influenced by a combination of many factors.

In the first six years of the Amboseli study, when the population was in decline, several things were happening. Low to average rainfall for three years, followed by drought for two years, resulted in very poor environmental conditions that were compounded by competition from over 10,000 Maasai cattle. In those years calf mortality was high, with over 33 percent dying before the end of their first year. There were also mortalities among older calves, especially around the age of

A group of elephants, including the T families, crosses an open pan on their morning journey to the swamps, with Kilimanjaro behind.

Drinking

Feeding on the coarse
swamp vegetation

A group rests bunched together with the smaller calves lying down under their mothers and older siblings.

Mudwallowing

A young calf "charges" the author's vehicle; some calves
are more adventurous than others from early on.

Slit Ear, Tara, and others feed on lush, abundant vegetation during a year of high rainfall.

A large male (not in musth) chases and catches a young female by placing his trunk along her back (note S-shaped curve of the penis).

She stops and he mounts, searching for her vulva with his mobile penis.

He achieves intromission by sinking down and hooking the penis into the vulva and then penetrating upward with most of his weight on his hind legs. (A young male watches in fascination but keeps his head very low to indicate lack of aggression and competition.)

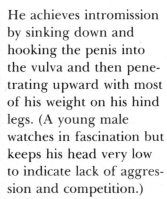

Males in musth have very swollen temporal glands which exude viscous liquid.

Greeting: Tallulah, Tuskless (center), Teresia (left), Tonie (right)

The "floppy run"

Tallulah and her newborn calf and "assistant" Tara: The baby, still very wobbly on its feet, reaches out and explores with its trunk (part of the umbilical cord still hangs down).

The calf tries to suckle on Tara's leg.

Tallulah and Tara have nearly buried the calf with fine soil and grass.

Tonie trying to lift her dead baby, who was either stillborn or died shortly after birth in November 1980 *Joyce Poole*

Phyllis and Keith with a
bushbaby that lived in
the camp

The author's tent and vehicle in Amboseli

Marc Hauser (left) and
a work crew hack
at the acacia tree that
fell on the author's
Land-Rover.
Joyce Poole

A family group feeds in the camp in the late afternoon.

Tom and Tao play in the lush grass as their older relatives feed.

The wary elephants from Kilimanjaro visit Amboseli for a few days at a time.

The Amboseli Maasai are changing their way of life: The traditional and the modern meet in a cabbage patch near Namalog swamp.

Ivory: It may be a beautiful substance when carved into an elegant figurine, but it was once the tusk of a living elephant who used it to feed and dig, play and fight.

weaning. In addition, there was poaching, spearing, and hunting, and thus adults were dying as well.

At the same time reproductive rates were low, showing the same kind of patterns that Laws had found in other elephant populations. The Amboseli females appeared to be delayed in reaching sexual maturity and were not giving birth to their first calves until they were 16 to 17 years old. The average calving interval was over six years.

A possible mechanism involved in reducing the birthrate was revealed by examining conception rates in those years of decline. From known births for 1976–78 I was able to estimate the conception rates for 1974–76. (Previous to this time my birth records were not accurate enough to estimate conception rates.) There were 24 conceptions in 1974, five in 1975, and none in 1976. To give a more accurate picture of reproductive activity, I worked out the percent conceiving of the total number of females available to conceive in each of those years. For this analysis females were considered available if they were over ten years old and were neither pregnant nor suckling a calf below the age of two years. The following figures were derived:

	1974	1975	1976
No. of females available	82	59	90
Percent who conceived	29%	8%	0%

It would appear that the Amboseli females virtually stopped reproductive activity for two years. As a result there were no births for 16 months from July 1977 to November 1978 and only two births for a period of 22 months from January 1, 1977, to November 24, 1978. The available females seemed to have stopped cycling. During 1975 and 1976 I saw only five females in estrus and none of them gave birth 22 months later. There is no evidence to show that females were cycling, conceiving, and then aborting, and elephants do not reabsorb fetuses under unfavorable conditions as some animals do. Human females and some other mammals stop cycling when body fat reserves go below a certain level and I suspect that this is what happened to the Amboseli elephants during the drought years. The elephants were not at the point of starvation (I had seen elephants dying from lack of food in Tsavo National Park a few years before) but they were in relatively poor condition and must have reached the critical threshold to stop the necessary hormone production.

Thus, in terms of population regulation in the Amboseli elephants, several things seem to come into play: high calf mortality, delayed puberty, long calving intervals, and cessation of reproductive activity altogether. Which of these factors are density-dependent and which environmentally induced is difficult to say. The elephants were at locally high densities during the drought, when they had to concentrate on the swamps, causing severe competition over the already poor and meager forage available there. The low quantity and quality of the food resulted in poor nutritional status, which led to high calf mortality and changes in female reproductive parameters and activity.

These records of the various changes that occurred during the drought were very important in understanding how elephants responded to unfavorable conditions. Just as important was how the elephants responded when conditions improved. After the drought finally broke at the end of 1976, the African savanna made its usual near-miraculous recovery, and soon there was an abundance of nutritious vegetation. It took only two months for females to start coming into estrus again, but not all the available females did so at once. Conceptions started taking place in Feburary 1977, reached a peak in July, slowed down during the latter part of the year, and then picked up again in early 1978, peaking that year in April during the rains. In all, 100 females conceived in the 19-month period covering February 1977 through August 1978. Needless to say this caused a baby boom approximately two years later, in 1978, 1979, and 1980. Five calves were born in November-December 1978, 57 in 1979, and 39 more in the first half of 1980, making 101 calves, the one hundredth birth being the twins that I have mentioned before. The population had increased by 21 percent in less than two years.

The pattern of resumption of reproductive activity was interesting to examine. The percent of females conceiving in 1977 and 1978 provided a contrast to the previous three years:

	1977	1978
No. of females available	116	101
Percent who conceived	36%	60%

However, in looking at the data I wondered why more females did not conceive in 1977. Only 42 of the 116 available females became pregnant. Among those 42 females an unexpected pattern emerged: Thirty-three belonged to central groups, three to peripheral eastern families that had immigrated and become central groups, and six from peripheral western families. The relative percentage of conceptions in each of the subpopulations was also significant: Forty-three percent of the available central females conceived, while only 19 percent of the available peripheral females did. It looked as though the females in the central families had recovered more quickly, possibly because they had not gone down so far in condition as the other elephants had. Even within the central families some females took longer to recover than others. Most of the females who conceived in the first two years were over 20 years old. The younger females who should have reached puberty needed a longer period to catch up. They and the peripheral elephants started conceiving at higher rates in 1979, two years after the others had resumed reproductive activity.

I had based my designation of "central" and "peripheral" on spatial sorting and not on differences in population structure, but when I discovered the delayed response of the peripheral elephants, I felt justified in saying that the Amboseli population could be divided into two subpopulations. There was a significant difference in reproductive success between the two and this opened up some important questions about how and why the two subpopulations are distributed the way they are, since it looks as if it is better to be a central elephant than a peripheral one.

At the same time that the baby boom was occurring, mortality rates were going down tremendously. The survivorship of first-year calves were very high compared to that in the drought years. Of the 101 calves that were born, 94 percent reached their first year, whereas only 48 percent of the 1976 calves had done so. The hunting ban, which had been instituted in 1977, and a successful crackdown on poaching reduced mortality among the adults. In addition the Maasai were happy with both the high rainfall and the agreement that had been made with the government, and so there was very little spearing as a form of protest. Even more important, in the following year, 1978, all the warriors, the *moran*, were promoted to junior elders and no moran was initiated for the next six years. It is the moran who have to prove their bravery by killing large, dangerous wild animals. Without a moran class the wildlife was left in relative peace.

Births and deaths are not the only data points that students of population dynamics are concerned with. In order to make predictions about what a population is going to do, one has to take other variables into account, such as sex ratios and age structures. I looked at these things as well in the Amboseli population.

The sexes of all but two of the 101 calves born during the 1978–80 baby boom were recorded. The two who were unrecorded died shortly after birth. The remaining 99 calves revealed a very neat 1:1 sex ratio as follows:

Year of Birth	Male	Female	Total
1978	4	1	5
1979	29	27	56
1980	17	21	38
Total	50	49	99

Previous and subsequent births also showed an overall 1:1 sex ratio, which seems to be consistent among the Amboseli elephants.

The age structure of an animal population is vital to know in order to make predictions about what will happen to the population in the future. For example, if the population has a large majority of young animals, then, when those individuals reach reproductive age, the population will almost certainly increase in the absence of high mortality caused by poaching or other external factors. On the other hand, if there are large numbers of middle-aged and old animals with few young animals to replace them, the population will probably decline.

In Amboseli, all the elephants born since the end of 1975 have known birth dates, accurate to within a month. For those born during the part-time study, 1972–75, the accuracy is within at the most six months and a majority are known to within three months. One- to three-year-old calves that were first seen in 1972 and 1973 could be aged fairly reliably by their size and tusk eruption and are considered known-aged to within a year. These calves bring recorded birth dates back to 1970. At the present time (1987), there are 360 animals from newborn to 17 years old whose ages are positively known, which is well over half the population.

These known-aged animals have made it possible for us to devise techniques to estimate elephant ages elsewhere in Africa. The most common aging method used for elephants involves the examination of their teeth. Elephants have six sets of teeth throughout their lifetime. Each tooth is a lozenge-shaped molar (the Latin name for the African elephant *Loxodonta* refers to its tooth shape) and there are only four molars in use at any one time, one each on the upper and lower two sides of the jaw. Each set of four new molars grows from the back and moves foward, pushing out the previous set. The timing of the eruption of the new sets of molars, when they come into use and when the old set falls out, is consistent with age. Elephants are born with molar sets 1 and 2 in the jaw; set 1 falls out when the elephant is about two years; set 2 is lost at about six years; set 3 at about 13 to 15 years; set 4 at 28 years; and set 5 at approximately 43 years. Molar set 6 appears at about 30 years of age and comes into use when the elephant is in its early forties. Very rarely elephants may have a seventh set of molars, but usually the sixth set is the final one and that has to last the elephant for another 20 or more years. By the time an elephant is in its sixties, its last set of molars is worn down. Eventually it cannot feed properly and death results.

Tooth eruption and wear is an excellent method of aging elephants but the major problem is that the elephant has to be dead (or captive and tame) in order to use it. Another aging technique is measuring shoulder height. The taller an elephant is the older it is. The method was first worked out by Richard

Laws, who took measurements of hundreds of elephants killed in culling schemes. He found a correlation between shoulder height and tooth age. Fortunately shoulder height can be measured on live elephants. Iain Douglas-Hamilton designed an ingenious device with two cameras, which took a double-image photograph of an elephant. This could then be measured for shoulder height. He was able to photograph most of the Manyara population and get fairly accurate ages for individuals under 20 and put the older animals in age classes. In Amboseli Harvey and I measured shoulder heights in a less technically sophisticated way with a graduated pole. First we photographed the elephant as it crossed a road or other level spot, then one of us would go with the pole to the very spot where the elephant had crossed and stand with the pole and be photographed. Later, when the two photos were printed, we could measure the shoulder height of the elephant against the pole.

A few years later Jonah, with the aid of his assistant Nick Georgiadis and Joyce and me, devised a clear, easy, and cheap technique for aging elephants. Jonah noted that the soil in Amboseli is soft and dusty for much of the year and that elephants leave very distinct footprints when they walk. He wondered if the size of an elephant's foot correlated with its shoulder height and thus its age. First we did an experiment. We took pole measurements of 30 elephants of various ages and sexes and at the same time measured their hind footprint. We chose the hind foot because elephants tend to place the hind foot on top of the forefoot print. On measuring the photos we discovered a very close correlation between shoulder height and footprint length. This meant that we simply had to measure footprints in order to estimate the ages of the Amboseli elephants.

We were then able to make the technique far more accurate by using our own known-aged animals. The original tooth and shoulder-height technique was based on only a few known-aged zoo animals. Amboseli was able to provide a much larger sample. In 1983 and 1984, when Phyllis was working on calf development, she measured the footprints of most of the calves and older juveniles and together with Keith was able to work out an age and growth curve which is probably the most accurate ever done for elephants. This new curve and the footprint-measuring technique have been helpful to scientists working on elephants in other areas of Africa. Jonah, Nick, and I published a paper in 1983 on the footprint technique entitled "Age estimation and population age structure of elephants from footprint dimensions" in the *Journal of Wildlife Management*. The growth curve was presented in the 1986 joint paper that Phyllis and I published in *Behavioural Ecology and Sociobiology* on "Early maternal investment in male and female African elephant calves."

However, these field-aging techniques on live elephants are still accurate only for animals under 20 years old. Since the rate of growth of females tapers off and nearly stops after 25 years old, the techniques cannot be used for assigning accurate ages to adult females. Males continue to grow at a perceptible rate but their age is not as important as that of females in population dynamics (in fact the males are left out of most models). The foot of an adult female grows less than an inch in perhaps ten years, making it difficult to distinguish a 25-year-old from a 35-year-old. For these older animals we have to estimate their ages and place them in classes of 5- or 10- or 15-year blocks.

To my bewilderment I have always been uncannily accurate at estimating the ages of the adult females by eye. In the early stages I was not totally conscious what I was cueing into, but whenever a female died and we collected her lower jaw to examine the teeth, the age I had assigned her by eye was always within two years of the tooth age. When I tried to sort out what I was doing, I realized that I took into account the size of the female, the circumference of her tusks at the lip, and her general overall appearance, how bony she was around the shoulders, how sunken her head above the eyes. Not only did I attempt to judge shoulder height, but with females I found it was more important to look at the length of the body. Old females, instead of getting taller, seemed to grow longer. Teresia, who came from a short-legged family anyway, had grown so long that she somewhat resembled a dachshund. People who were not on the elephant project knew Teresia for her funny shape and often remarked on it. Virginia, Jezebel, and Horatia, females in their fifties, also had long bodies, but being taller in the leg than Teresia, they did not look as odd.

One other important bit of information I used in estimating

the ages of the adult females was old photographs. The photographs taken by Norman Myers in 1964–66 proved to be extremely helpful. For example, Norman took several roles of film of the Ts. I was fascinated to see that way back in 1964, over 20 years ago, the same four families were together in a tight-knit coordinated group. I could recognize Teresia, Torn Ear, Slit Ear, and Tania, and I think even little Tuskless. With Torn Ear in the photos was a two- to three-year-old calf who should have had tusks by that age but did not. I am sure this was Tuskless. Teresia, Torn Ear, and Slit Ear were definitely big adult females even then. I estimated Teresia to be at least 40 years old in 1964 and put the others at ten or more years younger.

Each year, with accumulated knowledge, I was able to age the older animals more accurately. As the known-aged animals got older I had a better idea of what a 12-, 13- or 14-year-old elephant looked like. Then when I reexamined the early photographs, mine included, I was able to add on the relevant number of years and come up with an age at the present time. For example, I can now look at the pictures I took of Tallulah and Tess in 1973 and feel fairly confident that they were then about ten years old because today I can compare their size and tusk growth to an absolutely known-aged elephant of ten, Tabitha, from their own bond group. I have spent a great deal of time going back over my photographs, Norman Myers's, and Mark Boulton's, to refine the aging each year. In the meantime, as elephants died, we collected their lower jaws and aged them by their teeth, and in this way I was getting a better idea of what 30-, 40-, or 50-year-olds looked like. Joyce was using the same estimating techniques for the males, not so much for population dynamics as for the relationships between age and dominance and age and musth.

It is a peculiarity of elephant age structures that they almost always contain peaks and troughs—that is, there will be numerous individuals in one age class, very few in the next couple, and then a bunch more in the following class. In other words, there is not an even distribution of ages through the population. Looking at the structure of the known-aged female elephants in the Amboseli population in 1983, there are distinct humps for certain ages:

1983

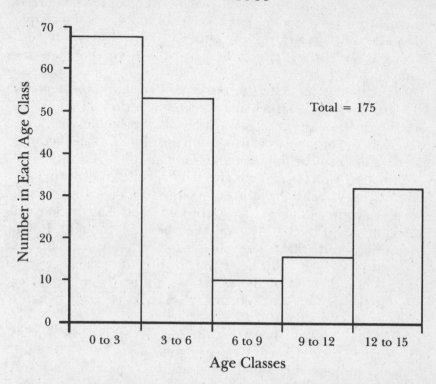

There were a lot of animals in the 12- to 15-year-old age bracket, few in the 9- to 12-year range, hardly any in the 6- to 9-year class, and then suddenly peaks of 3- to 6-year-olds and 0- to 3-year-olds.

I have already described one of the circumstances that cause these bumps to occur: Elephants can cease or be inhibited in reproductive activity under unfavorable conditions, and at the same time calf mortality may be high for those calves that are born. A second reason for peaks and troughs is an inherent biological constraint: Elephant gestation length and calving intervals make it difficult for there to be even age distributions. If conditions were uniform and continuously favorable, elephants might produce the same number of calves per year. If there were 100 females in a population and an average calving interval of four years, then one would expect 25 calves per year. It never seems to work this way. Once the population hits some sort of snag and the females stop breeding or there is high calf mortality, a wave pattern is set up. The females tend to breed in synchrony and produce many calves in the space of a couple of years, and because of the three- or four- or more-year calving interval they tend to breed again in synchrony a few years later.

To further accentuate the peaks and troughs, the females that are born in a particular baby-boom period all enter the breeding population at the same time and then create their own wave pattern, often in synchrony with the older females because they are affected by the same environmental conditions.

All these factors—the drought, the breeding synchrony, the calving interval, and the influx of a large cohort of young adult females—contributed to the unprecedented baby boom of 1983, during which 81 calves were born. I thought I had witnessed a remarkable baby boom in 1979–80 but that was over a two-year period. The 1983 baby boom basically covered only six months; 74 of the calves were born from February through July 1983, the remaining seven born from August through December.

What had happened was that the females who had calves in 1979 and 1980 were ready to give birth again in 1983. Thus the females who gave birth in 1979 conceived again in 1981 and gave birth in 1983 after a calving interval of about four years. The females who had given birth in 1980 surprised us by also conceiving in 1981 and giving birth in 1983 after only a three-year calving interval. The shortest interval I have recorded in Amboseli is two years and seven months, but I think this is unusual and one of the lowest possible. These intervals are for females whose previous calf was surviving. Shorter intervals are possible when a calf dies at or near birth. For example, Tonie, who lost her calf at birth, had a calving interval of only two years. The three-year calving intervals for females with surviving calves showed that elephants can step up their reproductive rates under favorable conditions, once again revealing the flexible responses of elephants.

The contribution of these mature females' calves to the baby boom was important, but what really made it outstanding was the coming of age of a whole group of young females who had been born in the years 1968–70. Forty-three young females reached sexual maturity around 1981 and many of them conceived in that year and gave birth in 1983. From the Amboseli records it appears that 11 to 12 years is the common age for puberty in females, although, as one would expect, there were a few who started cycling earlier, at nine and ten years old. At the same time there were some females who did not appear to give birth for the first time until they were over 20 years old.

By 1983 every adult female over 20 had given birth except for one, Paula. She had never been seen in estrus, nor had she shown signs of pregnancy. Paula was named Paula only recently because for many years I thought she was a young male. "He" was called "Petra's Young Male," and I had given "him" a male number, M176, and carried "his" photo identification card in the bull box. It was only when the young matriarch of the family, Petra, gave birth and I noted that M176 was paying a great deal of attention to the calf that I grew suspicious. My notes read:

> 1030 Petra's Young Male, M176, acts like a young female toward Petra's baby, backing in and reaching to touch it. This is the first time I have ever seen a male behave in this way.

> 1115 M176 is behaving in this way because "he" is a female!

I was shocked, as were Joyce, Keith, and Phyllis. Everyone thought she was a male. Her head and tusk shape were very masculine, but she definitely had female genitalia. However, the vulva was difficult to see because it was placed farther forward between the hind legs than is usual. I can also give myself another excuse for not being a total idiot since Paula and Petra are from the PB family, which is one of the most peripheral families and is not seen that often. In the early days they were frightened of cars and would not allow me to come near them. I simply never looked that closely at M176's genitals. But I think it is significant that Paula has never had a calf and I suspect that there is something abnormal with her hormones and possibly even with her reproductive organs. All the same, the fact that there is only one female who is abnormal shows that the Amboseli elephants are healthy and fertile.

One final question of interest in the study of population dynamics is when females slow down or stop reproducing. Females well into their forties are fertile and continue to produce calves regularly. There are only a few females estimated at over 50 years old in the Amboseli population, and although my data do not suggest a "menopause" in these animals, there are indications that the rate at which they produce calves is lowered and the ability to produce viable offspring may also be affected. Of the six or so females who are near or over 50, four of them—Teresia, Virginia, Lillian, and Big Tuskless—lost their most recent calves shortly after birth. As I mentioned before, Teresia and Virginia had not had calves for nine years before these last calves. After the calves died the mothers showed no further signs of conceiving again. This lowered fertility in the older age classes has to be taken into account when designing population models.

As far as drawing up a simulation model to try to say what the Amboseli population is going to do, I think it is mainly a theoretical and only partially valuable exercise. There is a danger of overrelying on the predictions of models. At the end of 1983, with the population at 674, it looked as if the elephants were rapidly increasing and would continue to do so, but, although models can factor in catastrophes and droughts, these events are still very unpredictable in their timing and severity. Even with 14 years of data I would still hesitate to try to say what is going to happen to the population in the next ten years. Conditions can change drastically. In 1983 no one could have known that 1984 was going to be such a disastrous year in so many ways for the Amboseli elephants.

In the meantime 1983 was a delightful time to be in Amboseli, both for the elephants and for the elephant researchers. Baby elephants truly appeared to be falling out of the sky during those first six months. It was wonderful for Phyllis, who had just started her study of calf development and was hoping she would have a large enough sample size. She turned out to have far more calves than she had ever hoped for. In fact she had all she could do on some days just to record the new births and get the sexes of the babies. By midyear there were over two hundred calves under the age of five years. She studied the young calves and also the behavior of their older siblings. The interactions between sisters and brothers was a fascinating mixture of sibling rivalry and care and concern.

The Ts continued to be a good bond group for the study of calves of various ages and combinations of siblings. Tania was the first of the Ts to produce a calf in 1983. She gave birth to a female around April 20. My records indicate that this was her first daughter in 20 years; she had had sons in 1962, 1968, 1973, and 1979. Slit Ear was the next to give birth, to yet another daughter, making this her sixth daughter in a row in the past 20 years. Tia had a new calf less than a week later, on May 23, and this time she had a daughter after producing three sons. It was about a month later that Tamar died in giving birth. Around July 4 Tallulah gave birth to her second daughter, and as far as we could tell (we did not see the actual birth) she took it all in her stride now that she was an experienced older female of 20 years. Finally, on August 22, little Tilly, Torn Ear's 13-year-old daughter, gave birth to her first calf, a male.

Despite Tamar's death it was an excellent year for the Ts in terms of reproductive success. It was to prove to be short-term, but at the end of 1983 all looked rosy—or, for elephants, the metaphor should be "green"—and they went into 1984 with the largest number in their bond group since I had first met them. Right Fang had become independent at 14 years old, leaving the four family units with the following 29 animals:

TA	Tuskless (62)	♀
	Tuo (80)	♂
	Tonie (67)	♀
	C'82	♂
	Tilly (70)	♀
	C'83	♂
TB	Tania (44)	♀
	C'83	♀
	Tom (79)	♂
	Toby (73)	♂
TC	Slit Ear (36)	♀
	C'83	♀
	Slo (80)	♀

Tabitha (76)   ♀
Tara (67)   ♀
  C'82   ♂
Tess (63)   ♀
  C'82 (50)   ♀

Tia (50)   ♀
  C'83   ♀
  Tio (80)   ♂
Taddeus (76)   ♂

TD   Teresia (22)   ♀
Tolstoy (71)   ♂
Theodora (67)   ♀
  C'82   ♂

Tallulah (63)   ♀
  C'83   ♀
Tao (80)   ♀

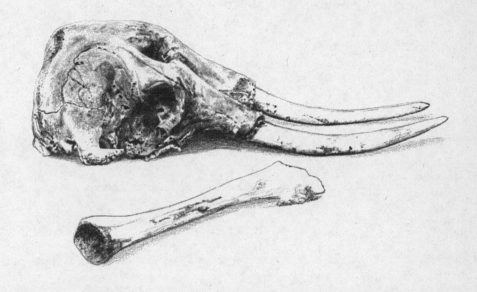

Chapter X

Life Cycle and Death

1984

The "long rains" of 1984 had been very poor, the lowest for many years, and conditions were approaching those of 1976. Rainfall had been below average in 1982 and above average in 1983 but that one good year was not enough to build up sufficient reserves, and when the rains virtually failed in 1984, there was another full-fledged drought. In June food was already scarce.

The Ts had migrated out of the basin briefly during the April and May rains, but they were soon forced back to the Amboseli swamps when the temporary water holes dried up and the ephemeral flush of grass gave out. As always, the TAs and TBs, led by Tuskless and Tania, returned to their usual haunts around the lodges and bandas. Tuskless had given birth to a calf in March, and the renewed demands of lactation increased her need for nutritious food. She led her family on a continuous round of the rubbish pits and even pilfered the banda's garbage pails and stuck her trunk right into the kitchens, totally terrifying the staff and tourists. In the evenings the TAs and

TBs crossed over to the Serena Lodge, situated at the source of the Enkongo Narok swamp, and broke into its garden area to feed on the decorative plants, some of which were quite tasty. Serena's garbage pit was also just a slight depression in the ground and there was no problem foraging there.

The TCs and TDs, led by Slit Ear and Teresia, started the dry season with their usual routine of spending the night in the woodlands and bushlands to the east and south and coming into either Longinye swamp or the small swamps in the Ol Tukai Orok area in the daytime. However, as in 1982, they found many other elephants using these areas and not very much food to go around.

In mid-June Slit Ear initiated the strategy that Teresia had used in 1982 by leading the two families to Namalog. Teresia was now over 60 years old and left most decisions about daily movements and activities up to Slit Ear. She always followed, making her way in her own good time. She often came along 30 yards or so behind the others, not limping or stiff or ill, but just very slow. Her teeth were worn down to small smooth discs about one quarter the size of the original molars. This was her sixth set of teeth and no new ones were coming in. She found it difficult to chew the coarse, fibrous swamp vegetation, and the tough bark and hard branches of trees were even more difficult to process. Still, with extra chewing and longer feeding bouts, she was able to get enough to eat when there was food available.

On the night of their move they had come out of Longinye and had fed and then slept for a while in the Olodo Are woodlands. When they got up, earlier than usual, at about 3:00 A.M., Slit Ear turned toward the east and after some gentle rumbling, foot swinging, and coordinating of group members, she headed toward the park boundary and Namalog. Teresia concurred and followed along.

The TCs and TDs together were quite a respectable group size now, and walking along in a column of about two abreast, they were 19 animals. Slit Ear, with her long, sweeping, upcurved tusks gleaming in the moonlight, was in the lead. She was followed closely by her year-old calf, three more daughters, and her grandson, Tara's C'82. Then came Tia and her three calves—one daughter and two sons. (Right Fang sometimes still joined his family or the TAs and TBs, but mostly he fed around the camp in the big glade that he knew so well.) Next came Tess and her daughter, and after a slight gap Tallulah came along leading the TDs. She was followed by her two daughters, and Theodora and her son, and Teresia's big son Tolstoy. There was another break of about 20 yards, and then slowly but steadily came old Teresia, with her two broken-off tusks and her long, low, swaybacked body. She was as dignified as ever.

The two families walked deliberately but not rapidly and left the park well before dawn. They passed the Maasai wells and two enkangs and moved on to the tip end of the Namalog drainage. After crossing the depression, they skirted along the northern edge of the swamp until they entered a stand of tall *Acacia xanthophloea* trees. Here the elephants fed for a while before continuing on along the swamp edge. However, they soon came upon something they rarely encountered in their range—a fence with posts and wire, not a thorn barrier which the Maasai made around their living areas. They knew the fences around sections of the lodges and they had seen fences on the southern side of Namalog several years before, but there were none anywhere else in the areas they went to. When they had spent the two months at Namalog in 1982, there had been no fences on this side of the swamp, nor had there been people this far west. From beyond this fence came the strong smells of Maasai and domestic stock and an unpleasant odor of strange vegetation. Slit Ear and the others turned back and entered the deep swamp, crossing to the island where they had been left in peace before.

When day broke, they heard the sounds of cattle bells and Maasai whistles and shouts, and later the chopping and breaking of wood and a thumping on the earth. All this commotion made the elephants uneasy, especially because it came from close to the swamp edge and not that far from the island. Now that they were there, the elephants stayed on the island for the day and left after dark, retreating back into the park for the night. The food had been fairly abundant on the island and in the swamp around it, and the next morning they started back to feed at Namalog again. Once more they entered the deep swamp just before dawn and fed for most of the day, with only the disturbance of the sounds coming across to them.

However, around 4:30 in the afternoon, a mixed herd of sheep and goats accompanied by some Maasai boys came down to the swamp edge directly across from the Ts. Some of the goats started into the swamp, hopping from one clump of vegetation to the next. All the elephants' trunks went up like periscopes, reaching out of the reeds and papyrus. The wind was coming directly from the boys and the livestock to them, and the intruders were too close for comfort. The elephants started to hurry out of the swamp, heading for the opposite shore to the southwest. They plunged through the water and mud and thick entangling roots. The adults simply broke through the vegetation, but the calves kept getting stuck. Theo's C'82 got caught in a mass of roots and screamed in fright as he started to go under the water. Both Theo and Teresia came to his rescue and helped him get free. All the elephants got to the shore, clambered out, and began walking rapidly, almost at a run, toward the west and the park.

It was the wrong time to be leaving the protection of the swamp, but they had had no choice. They kept to the edge of the swamp, although on this side it was fairly open. They had almost reached the end of the Namalog drainage when suddenly, from behind a clump of young acacias, stepped three Maasai men, each with a long glinting spear. Slit Ear saw them first and gave an alarm, and they all turned and ran. The three Maasai sprinted toward the elephants at an angle, trying to cut Slit Ear off, but she swerved and got past with her family and Tolstoy, Tallulah, and Theo and their calves closely bunched behind. Out on her own, unable to run as fast as the others, was Teresia.

The men quickly raced toward her with spears poised in their right hands, raised parallel to the ground and above their shoulders. The first man threw his with a tremendous thrust combined with the momentum of his sprint. It hit Teresia in the left shoulder and drove deep into the muscle more than six inches. The next man flung his spear just as she turned to charge. It went through her ear and into her neck. The third man hesitated, aimed carefully, and with all his power threw his spear and hit Teresia behind the shoulder. He had aimed for the heart and had just missed. The blade went slightly to the right and between two ribs.

Teresia groaned in pain with each strike and staggered when the third spear hit, but she did not fall down. She charged, and with no more weapons the Maasai had to run. After chasing them only briefly she turned back toward Namalog and in agonizing pain made her way around to the northern side. Two of the spears had fallen out, but the first, the one that had gone into her shoulder, was still embedded. It scraped against the bone as she tried to run. She limped her way to the thickest part of the swamp edge and hid among the dense bush. There was no way she could catch up to the others, who were now halfway back to Olodo Are, running at full speed and very frightened.

When Teresia reached the thicket she stopped, her sides heaving with the effort. With her trunk she grasped the spear and pulled and twisted it until it came out. The original penetration had torn through her muscles and in wrenching it out she did more damage. She had made it to the thicket by sheer will. Now she could barely move. The three men had not tried to follow her. For the moment she was safe, but they or others could find her later and finish her off with more spears.

However, Teresia had had much experience in avoiding Maasai during her life and she managed to elude them by sticking to the thick bush and keeping just ahead of the herders when they came near. She was in considerable pain and was seriously crippled by her shoulder wound, but she was not near death because the spears had not severed any arteries, nor had they hit any vital organs. After two days of hiding while recovering her strength, she was able to move down to the swamp to drink and even feed for a while. She was not strong enough to try to make it back to the park and her family. She had to remain at Namalog on her own. A few Maasai saw her during this period and noticed that she was wounded but they were not interested in killing her. It was not very exciting or brave to kill an old wounded animal, so they left her alone. For her attackers it had been a spur-of-the-moment sport.

The danger to Teresia did not come from the Maasai, but in a more insidious form. Teresia's wounds were narrow and deep, which was very bad for an elephant. The thick, tough skin tends to nearly close up around the openings, making it difficult for puncture wounds to drain properly. Teresia's shoulder wound was draining well because it was kept open by her movement. The other two wounds were not draining and soon became infected. It was a slow process, but eventually the infection spread and turned to septicemia, a general poisoning of the blood. All three wounds oozed a greenish-yellow pus.

Ten days after she was speared, Teresia was still alive but losing the battle with the infection, which had spread throughout her body. She was in less pain now but feverish; she barely fed and just managed to get to the swamp edge to drink. She was thirsty most of the time because of the fever and so tried to stay as close to running water as possible without meeting Maasai.

After two more days Teresia could no longer move and stood in the shade of a tree, swaying slightly with unsteadiness and only semiconscious. She had reached a state of feeling little and she mostly dreamed, perhaps of vast swards of sweet new grass and clear, cool hill streams. Or the taste and feel of the sweet juices that squirted out when she crunched down on her favorite wild fruit. Or most likely the smell and touch and sounds of her family—those who were still alive and others who were long gone, her grandmother, mother, sisters and brothers, sons and daughters.

On her final day Teresia grew too weak to stand and slowly subsided onto the ground, first onto her chest and then over on her side. She was conscious of the sounds and smells around her for a while longer, but soon there was nothing. Her long life was over.

I am not going to try to pretend that I was not more affected by the death of Teresia than by that of any other elephant. I had grown very attached to that funny-looking old elephant and I knew I would feel sorrow and loss when she died. By the end of 1983, when I saw how slow she was and how old she looked I knew she was going to die in the next couple of years of natural causes if nothing else because of her age and her teeth. But I did not want Teresia to die in pain and suffering and I worried about her. Each time I went to Amboseli or each time Phyllis and Keith came to Nairobi, I asked about Teresia in particular, not just "Have there been any deaths?" but "Have you seen Teresia? Is she all right?"

When I arrived in Amboseli for a short stay in June 1984,

Phyllis and Keith reluctantly told me that they thought "something" had happened to Teresia. They knew it would upset me but they had to tell me. Phyllis had found the TCs and TDs running through Olodo Are with their tails up and their ears tilted up and back in a posture of fear and distress. Teresia was not with them. During the following week the two families were seen again, and Teresia was still not with them. We were fairly sure she was ill, wounded, or dead and I went back to Nairobi feeling frustrated and unhappy about not knowing what had happened and where she was. Then the next week I got a message telling me that Teresia had been seen at Namalog by a Maasai friend of ours, Daniel Nkurupe. She had been speared and was ill. Daniel had gone out elephant watching with Joyce many times and knew several of the elephants by sight, including Teresia, who with her odd shape and broken tusks was very easy to recognize. The news that she was alive gave me some hope but I was distressed to think of her alone and wounded at Namalog. A couple of more weeks went by and Daniel did not see Teresia again, nor did she return to Amboseli. We knew she was dead.

However sad we all were, we also did not want her death to go to waste in terms of scientific information. She was estimated by me to be the oldest elephant in the population and we very much wanted to collect her lower jaw to see how old she was by the tooth aging method. Keith and Phyllis set to work as detectives. All the Maasai in the area knew who had speared Teresia, but no one would say who they were and no one would say where the carcass was in case it was assumed that that person had been responsible for her death. After much negotiation in a covert operation involving Maasai friends and intermediaries, a Maasai led Keith and Phyllis to the carcass. They found the partially rotted body and saw that Teresia had spear wounds but not many. The tusks were gone. They worked loose the lower jaw and the upper molars and took them back to camp. By Laws's aging criteria Teresia was 62, plus or minus two years. I had estimated her to be 60 and had put her year of birth at 1924. I now revised it to 1922.

It was intriguing to try to imagine what Teresia must have experienced in that long life. For half of her life she probably very rarely saw cars and encountered few man-made structures other than Maasai mud and dung huts. There were only elephants, other wild animals, the Maasai, their livestock, and sometimes people who came to hunt or poach. When she was well into her thirties, things began to change and more vehicles appeared and buildings went up in places where she used to feed. The cars probably frightened her for many years but eventually, when nothing bad seemed to occur around them, she got used to them. In the last ten years of her life she tolerated them completely and almost certainly knew by sight and smell the research vehicles with the particular people in them that followed her for hours at a time.

Changes in Teresia's habitat caused by conservation efforts and tourism may have been inconvenient and possibly irritating at times, but I suspect that they were basically extraneous and unimportant to her life in general. I think what mattered most to Teresia was her family—her calves and her other relatives; and her concern was how best to protect them and secure the resources that she and they needed to survive and flourish. I like to think that Teresia was able to pursue these goals without too much interference and that she had a relatively good life for an African elephant living in the twentieth century. Very few elephants are able to live out their full lifespan in the way that she did. Amboseli is probably one of the last places in Africa where she could have. Elephants have been pushed out of so many places where they used to be found and squeezed into small sanctuaries. To have to break with tradition, abandon well-known home ranges, give up old migration routes, and crowd into a restricted area must be stressful. Both inside and outside parks and reserves, most African elephants are pursued for their ivory and the older elephants are always the first to be killed because their tusks are bigger. In the few areas where poaching is under control, elephants are culled for a variety of reasons, usually "for their own good." Whole family units are killed from the matriarch down to the littlest calves, with only a few calves between one and three years old saved to be sold to zoos. As painful and agonizing as her death was, I would still prefer that Teresia died in the way that she did rather than be killed "cleanly, efficiently, and painlessly by a bullet" in a culling scheme in which she would have lost her whole family, all her female descendants, most of her genes, nearly everything she had lived 62 years for.

Looking back over Teresia's life provides an opportunity to review the full life cycle of a female elephant. When Teresia was born around 1922 she weighed about 260 pounds and stood approximately two feet nine inches at the shoulder. Her brain weighed 35 percent of what it would weigh as an adult. (The human brain is 26 percent of its adult weight at birth; most other mammal brains are 90 percent of the adult weight at birth.) She had a lot of mental and physical development ahead of her.

Teresia probably spent the first four years of her life in very close proximity to her mother and her siblings and other relatives in the family. She depended on her mother for both milk and care. When she was around four years old, a younger sister or brother was most likely born and Teresia was weaned. At four she would have taken great interest in her new sibling and lavished it with attention. At the same time she was developing her own social relationships within the family and learning elephant protocol. As a young calf she would have been tolerated in most situations, but as she got older she had to learn about dominance and aggression as well as affiliation. She would also have begun to know elephants from other families and her own family's position in relation to them.

The years from 4 to 12 or 13 were probably by elephant standards fairly carefree. Teresia was growing, learning, playing, and developing friendships and bonds. I am sure there were many bright-green, sun-filled days when Teresia and the other youngsters in her family and bond group were able to "be silly." I can imagine them racing about, beating through bushes and tall grass, heads up, ears out, eyes open wide glinting with mischief, or running across the open pans in the "floppy run," letting forth wild, pulsating play trumpets. On the best days everyone joined in, even the adults.

Starting sometime in her ninth or tenth year Teresia's body began to go through changes and in her eleventh or twelfth year, if conditions were adequate and there was not a drought, she probably came into estrus for the first time and experienced all the confusion, fear, and excitement of that event. She may or may not have conceived on that occasion. If she did not she probably came into estrus again a few months later and most likely conceived then. Her body went through further changes as the fetus grew and various hormones went into action. Her breasts filled out gradually over the months. By the time she was 18 months pregnant her sides were bulging and the weight of the fetus slowed her down. She probably had to feed more than before, because she, as well as the fetus, was continuing to grow. At her age she was only about two thirds the size of a full-grown adult female.

At 13 to 14 years old Teresia probably gave birth to her first calf. Most likely she would not be in synchrony with the other females in the family and she may also have given birth out of the main birth season. Given the odds for first calves, this one may not have survived, especially if it was a marginal year in terms of rainfall. But in a good year and in a close-knit family, it could very well have lived beyond its first year. Born around 1935, it would have been 49 the year Teresia died. She had no living daughter that old and there were only one or two big bulls of that age. I suspect that animal died somewhere along the way.

If her first calf lived out its early years, then Teresia would have come into estrus again when it was about two and given birth to a new calf when it was four and she herself was 17 to 18. For the next three decades of her life Teresia was either lactating or pregnant or both, as she would continue to suckle her youngest calf until the next one was born. If she gave birth to her first calf when she was 14 and had a calf every four years until she was 49, when Tolstoy was born, she could conceivably have had ten surviving calves. If she had her first calf at 12 and had a few very short calving intervals of only three years, then she might have had 11 calves in those years. After 50 Teresia's breeding slowed down considerably. We know that she did not produce a calf for nine years, and then gave birth to her final calf when she was 58. That one did not survive.

In those middle years, when she was reproductively most active, important social changes were also occurring. Whoever was matriarch when she was born, possibly her grandmother, would have died and the next-oldest female taken over. This may have happened two or three times before Teresia herself was the oldest female and therefore the matriarch. I know that she was the oldest female in 1964 because among the photographs that Norman Myers took there were several complete rolls of film on the Ts. They were standing in a group resting under some trees and he photographed them from all angles. Teresia was easily recognizable and she was definitely the biggest female there. Thus she must have been matriarch of her family and the senior member of the whole bond group for at least 20 years.

I think it could be said that in the last decade of her life Teresia was in her old age. I do not know if elephants become senile. Teresia always seemed to have all her faculties intact. She stayed with her family, continued to interact with them, and was as tolerant and sweet-natured as she was when I first knew her. Teresia must have been a particularly caring and in some way attractive grandmother because Theodora's calf spent far more time with her than with his own mother. In those last two years, whenever I saw the family, he was often following Teresia or standing next to her. He would go to Theo to suckle but then would return to Teresia. If I had not known the family I might have thought it was her calf. She also continued to maintain a close relationship with her strapping son Tolstoy, who was 13 years old and showed no signs of going independent up till the time she died. Thus, although slow and possibly not leading the family in the way she must have when she was younger, Teresia was still a vital part of the family in her old age.

I would estimate that 65 is probably about the oldest age an elephant can reach in the wild. By then the teeth would be completely worn down and malnutrition and its related diseases would follow. The famous Kenya elephant Ahmed of Marsabit mountain died in this way, although he was only 55; but he had an abscess under the molar on one side of his jaw and he could chew only on the other side. The molars on his chewing side were worn away to nearly nothing, while the other two were untouched. I saw pictures of him before he died; he was gaunt and emaciated. Postmortem examinations revealed no disease other than the abscess. He had probably died of starvation.

Very few elephants live out their full lifespan. The single greatest cause of death is killing by man, whether it be poaching, hunting, or culling. Natural mortality results from a variety of causes. As I have shown, mortality can be high among young calves in drought years. Again, this type of mortality is linked to nutrition. Once a calf becomes about five years old its chances of survival are much higher. There then seems to be a second small rise in mortality around the age of independence for males. We do not know how they die, but it seems to be a time of vulnerability for them. They may be more likely to be speared or poached, but in Amboseli we do not have direct evidence that this is occurring, only that young males are dying at a higher rate than would be expected. After this age natural mortality levels out to about 3 percent per year for adults.

In Amboseli most elephant deaths are a result of drought or spearing, but some animals have died of other causes. We know of several accidents that have resulted in death: breaking of legs and falling into pits and wells. And Joyce saw one of the Amboseli elephants, Polly, fall over and die, and the sole thing she could find was two bleeding puncture wounds on one of her legs. She could only conclude that a snake had bitten her. It must have been a large puff adder, black mamba, or cobra to get through the skin and inject enough venom to kill her. Other elephants have appeared to die of disease. We have never had a veterinarian working on the project and have no facilities for doing postmortem examinations (a formidable task with an elephant) and thus we cannot explain the deaths of some of the elephants. Anthrax has been suggested, as has blocked intestines.

In areas where dead elephants have been examined, a variety of diseases have been found. Abscesses of the tusks, teeth, and jaw were frequent ailments. Sylvia Sikes, working on culled elephants in Uganda, found cardiovascular disease in adults. Calcification of the blood vessels was the most common condition. There were also fatty deposits in the arteries. Both conditions could lead to heart attacks or strokes. Working elephants in Asia are known to die from both.

Elephants seem to suffer rarely from any contagious epidemic diseases such as rinderpest, which has a devastating effect on the cloven-hoofed animals. However, in Lake Manyara National Park, a disease did sweep through the elephant population, killing over 100 animals. It was thought to be pneumonia-related. Anthrax is a contagious disease, but usually only one or two elephants succumb and only those in poor condition already. Rabies has been recorded in Asian elephants, as have foot and mouth disease, tetanus, tuberculosis, pneumonia, dysentery, and something called elephant pox, which is related to chicken pox. All these diseases may be a result of domestication.

In Amboseli elephants sometimes suffer from a condition I have called the "twisting disease." It usually occurs in the wet season to one or two elephants at a time, and it does not seem to spread through a family. It affects calves over the age of about five, adult females, and males up to about 25 years old. The symptoms are a general stiffness of the limbs, an odd carriage of the head, and a bizarre way of walking. The malady causes the body to twist as the elephant moves. As far as I know, elephants do not die of twisting disease. They seem to get over it within a few weeks and then return to normal. I have always suspected that the condition is caused by a virus that affects motor coordination. On the other hand, it may be caused by some plant that these individuals have eaten. Many plants have toxins of one sort or another and can only be ingested in small quantities.

Blood tests have shown that African elephants carry a variety of parasitic diseases such as trypanosomiasis and tick fever, both caused by protozoa. Microfilariae parasites have also been detected in the blood, but it is not known what diseases they may cause. Various other internal and external parasites are common in elephants. They are hosts to several different kinds of worms in the guts, stomach, lungs, and bile ducts. Elephants also suffer from ticks and botflies. None of these parasites seems to kill elephants directly, although they could hasten the decline of an elephant that is ill from something else.

While elephants that are ill do not go off to an elephant graveyard to die, the myth is not without some foundation. Probably in any given elephant range there are places that sick and wounded elephants tend to go. These would be areas where there is water and shade and soft vegetation to eat. In Amboseli sick elephants often come to Ol Tukai Orok, the woodland where my camp is situated. Dotted about this woodland are small swamps. Elephants who can no longer travel stay close to these swamps in the shade of the big *Acacia xanthophloea* trees. Places such as this might have more carcasses than other parts of a population's range, and therefore people may have thought there was a special area where elephants went to die. More likely the myth arose from hunting practices that involved killing several elephants at the same time. One tribe in southern Africa used to kill whole herds of elephants by building a ring of fire around them. In this case the bones of many elephants would have been found in one place.

Elephants may not have a graveyard but they seem to have some concept of death. It is probably the single strangest thing about them. Unlike other animals, elephants recognize one of their own carcasses or skeletons. Although they pay no attention to the remains of other species, they always react to the body of a dead elephant. I have been with elephant families many times when this has happened. When they come upon an elephant carcass they stop and become quiet and yet tense in a different way from anything I have seen in other situations. First they reach their trunks toward the body to smell it, and then they approach slowly and cautiously and begin to touch the bones, sometimes lifting them and turning them with their feet and trunks. They seem particularly interested in the head and tusks. They run their trunk tips along the tusks and lower jaw and feel in all the crevices and hollows in the skull. I would guess they are trying to recognize the individual.

On one occasion I came upon the carcass of a young female who had been ill for many weeks. Just as I found her, the EB family, led by Echo, came into the same clearing. They stopped, became tense and very quiet, and then nervously approached. They smelled and felt the carcass and began to kick at the ground around it, digging up the dirt and putting it on the body. A few others broke off branches and palm fronds and brought them back and placed them on the carcass. At that point the warden circled overhead and dived down in his plane to guide the rangers on the ground to the dead elephant so that they could recover the tusks. The EBs were frightened by the plane and ran off. I think if they had not been disturbed they would have nearly buried the body.

Even bare, bleached old elephant bones will stop a group if they have not seen them before. It is so predictable that filmmakers have been able to get shots of elephants inspecting skeletons by bringing the bones from one place and putting them in a new spot near an elephant pathway or a water hole.

Inevitably the living elephants will feel and move the bones around, sometimes picking them up and carrying them away for quite some distance before dropping them. It is a haunting and touching sight and I have no idea why they do it.

When an elephant dies in Amboseli we let it rot for a while and then collect the lower jaw for aging. The jaws are often still smelly so we put them out in the sun on the periphery of the camp. Without fail these hold a fascination to all passing elephants. Recently one of the big adult females in the population died of natural causes and we collected her jaw after a few weeks and brought it to the camp. Three days later her family happened to be passing through camp and when they smelled the jaw they detoured from their path to inspect it. One individual stayed for a long time after the others had gone, repeatedly feeling and stroking the jaw and turning it with his foot and trunk. He was the dead elephant's seven-year-old son, her youngest calf. I felt sure that he recognized it as his mother's.

I have often wondered if elephants experience anything akin to grief when a close family member dies. Certainly the death of an important animal such as the matriarch has a profound effect on the family. There can be total disintegration of the group for a long period afterward, with some families never resuming their former cohesiveness. I have seen females whose young calves have died looking lethargic for many days afterward, sometimes trailing along way behind their family. They had shown no signs of illness before the calves died so it may be that they were "depressed" in some elephantine way.

Unfortunately for all of us, elephants and researchers alike, 1984 proved to be a year in which the themes of drought, disease, death, disaster, and depression were paramount. If 1976 had been a bad year for the project, 1984 far surpassed it. To start with and underlying many of the problems, there was the total failure of the long rains. To compound the difficulties severe financial cutbacks in the Wildlife Department meant that the park did not have the money to pump water out to the Maasai, nor could they maintain the boreholes. There was no choice: The Maasai were allowed to come into the park for water and eventually, as the drought intensified, for grazing as well.

As in 1976, competition for the remaining resources was acute. To aggravate matters the Maasai had begun to circumcise their young men and initiate them into the warrior class as moran, which meant that incidents of spearing increased. The combination of factors spelled disaster for the elephants. In 1984, 67 elephants died, a number second only to the mortalities of 1976. The deaths started slowly with the loss of two of the '83 calves in February and March, both the first calves of young females, making their deaths not unusual. One elephant died in May, three in June, six in July, and one in August. But then the drought began to take its toll and in the next four months over 50 elephants died. Many of them were calves who suffered poor nutrition or the loss of their mothers, but there were also many adults who were speared, with only a few dying of natural causes. In all, 11 adult females, 13 adult males, 3 juveniles, 13 weanlings, five second-year calves, and 22 first-year calves died during 1984.

For once I did not envy Phyllis and Keith being in the field while I was writing in Nairobi. It was a very difficult year for them. They were the ones who kept finding dead, sick, or speared elephants. Phyllis had to watch her babies grow thinner and thinner and one by one disappear. She and Keith were the ones who had to console their Maasai friends and help them with deliveries of beans and maize meal. (In parts of Kajiado district, Maasai children were dying and one of Kenya's aid organizations had to set up a relief center.) They even collected money to buy the diesel to pump water to the Maasai who were up on the ridges, and they tried to get spare parts for the boreholes. But as the drought progressed, their efforts seemed hopeless.

I remembered how on edge I had felt during the 1976 drought, and I worried about Phyllis and Keith, who were both of a volatile nature anyway. They came through it without breaking down, but Phyllis said she came close to it the day she found little Fritz, Freda's five-year-old son, dead and looking like a macabre pin cushion from being speared over and over again in what must have been a gruesome massacre by a gang of moran. Keith held his temper most of the time, although on one occasion he vented all his frustration and anger at a tour leader who had driven his vehicle to the top of Observation

Hill, at the base of which, in order to prevent soil erosion, there was a sign asking people not to drive up. Seeing the car up there made him snap and he was outrageously abusive to the tour leader, who in turn got exceedingly heated himself. When I heard about the incident I sided with Keith completely and understood exactly why something like that would make him so angry. When one cares a great deal about a place—its animals and its habitats—and one witnesses it undergoing such terrible changes during a drought, the added insult of human destruction by someone who should know better can easily set one off.

In the meantime, I was not immune from the disasters of 1984, even in Nairobi. A whole series of difficult and unpleasant incidents occurred. It was definitely a year that made one reassess one's work and life in Africa. However, I never seriously considered leaving. I cared too much about the project and the elephants. In fact, in strictly scientific terms, 1984 was a valuable year. We learned a great deal about mortality patterns in elephants. Keith collected important data on elephant feeding ecology under drought conditions, in contrast to his earlier study under favorable conditions. Phyllis was able to assess various crucial factors in calf development and survivorship. And I confirmed my 1976 findings on the effects of deaths and droughts on the behavior and social relationships of elephants.

One of the things I was interested in was how individual families fared during the drought. In 1976 some families obviously suffered more than others, and I wondered what factors were involved, whether it was the area they lived in or their status in a possible dominance hierarchy among the families, or a combination of both. The results were not clear-cut and they were often contradictory. In 1976 the Ts came through the drought with no losses at all. I speculated that they had done well because they lived in one of the richest parts of the park and because they probably benefited from the leadership of their older matriarchs. In 1984 the T bond group was one of the hardest hit of all.

It started in May with the disappearance of Tania. During the briefest of rains Keith found the Ts over in the western part of the park and Tania was not with them. On the next sighting Phyllis found Tania's '83 calf, plus Tom and Toby, with

Tuskless, but there was still no sign of Tania. Keith and Phyllis began to search for Tania's carcass both on the ground and from the air, but with no luck. The '83 calf, who was just over a year old, stayed with the TAs but Phyllis sadly watched her grow weaker and weaker until she died in July. Around this time they finally found Tania's carcass. She was only a couple hundred yards from the main lodge garbage pit. Since the carcass was over two months old, it was difficult to see if there had been any wounds. We thought it was much more likely that she had eaten something in the garbage that killed her.

Needless to say, we found this very upsetting and continued to try to get the lodge and the Kajiado County Council to make their garbage pits elephant-proof. Keith began to collect all the things we found in elephant dung: plastic bags and gloves, medicine bottles, broken glass, bits of metal, and various kinds of wrappings and containers. It was distressing to find these things littered over the park and to know that they had passed through the digestive tract of the elephants. He hoped to appeal to the various people in charge by showing them these things, but no one seemed to care. Eventually all the researchers in Amboseli chipped in to have the smaller Ol Tukai garbage pits rebuilt with steep sides and wire grills on top like the pit we had in our camp. With this system the baboons and vervet monkeys were also kept out. Keith designed the pits, organized the labor, bought the grill work in Nairobi, and transported it to Amboseli. The situation improved greatly. Unfortunately the lodge garbage pit was too big to rebuild.

So Tania and her C'83 died from what we called "garbage-related causes" on our mortality lists. In the meantime, Teresia had been speared at Namalog in June, making three losses among the Ts. It was not to end there. The drought began to be severe in September and got worse and worse through October and into November. The spearings also stepped up with so many Maasai in the park, plus the new moran. Not only were adults being speared but calves as well. Calves who had been weaned and were weakened by the drought, as Audrey had been in 1976, left their families and stayed near the swamps. Whenever we went out we now found calves on their own. They were an easy target for moran with nothing better to do. We lost twelve calves who were four to five years old.

Significantly, all but two of them were males, further substantiating Phyllis's theories on differences in survivorship between sexes. Males, needing more food because of their faster growth rate, probably weakened sooner than females and separated from their families. They then either were speared or died of malnutrition.

Some elephants also appeared to die of disease during the drought, probably through illness brought on by their poor condition. In October Keith and Phyllis came across Tia with just her C'83 in the Ol Tukai Orok woodlands. She looked ill and lethargic. They saw her a few more times still on her own, and on October 17 they found her dead and her calf gone. The calf was never seen again and must have died shortly after she did.

Once again I was very disturbed to hear this news. I had closely followed the lives of the Ts for 11 years and I felt the loss of Teresia, Tania, and Tia intensely. I also knew the families would never be the same again. First of all, with the death of Tania the TBs no longer existed as a family. Tania left only her two sons Toby and Tom and no female to carry on. Toby was ten and old enough to be independent but Tom was only five and at a vulnerable age. After Tania's death Tom wandered around either on his own, with the TAs, with other family units, or with other calves who had separated from their families. He had been weaned in 1983 when Tania had her new calf, but I worried about him nonetheless, given the record of young males his age.

I was also concerned about Tio, who was only four and a half. He too had been weaned, but he seemed far less independent than Tom. Tia's older son Tad was eight years old and I thought he would be all right.

At the beginning of November Phyllis and Keith came to the end of their projects and left Amboseli with a mixture of sadness and relief. The drought was still in progress and they were strung out and tired from their work and from the whole situation in Amboseli. At the same time they did not want to leave the elephants, whose lives they had become so wrapped up in. Joyce was starting a new project on the bulls and also one on elephant vocal communication. She would take over some of the monitoring until Sandy Andelman came to work on the cows and calves after finishing her Ph.D. Sandy, like Phyllis, had been lured away from the vervets by a fascination with elephants. Meanwhile I tried to spend more time in Amboseli to keep up with the deaths that were still occurring.

After Phyllis and Keith left, Joyce and I recorded 36 deaths in November and December. Some of them, particularly the males that Joyce found missing, probably occurred earlier in the year, but the majority of them took place in those final months. Among the animals that died then were Tio *and* Tad. Phyllis saw both Tio and Tad on November 1, just before she left. I saw Tio once in November with the EBs, but after that he was not seen again. I never saw Tad after Tia's death and I do not understand why he died. He was eight years old; he was not dependent on his mother for nutrition. He probably could have stayed with the rest of the TCs, or possibly he could have joined his older brother Right Fang. I have to suspect that Tia's death affected him in such a way as to make him more vulnerable to the drought.

Another factor may also have increased Tad's chances of dying after the loss of his mother. He was born in 1976. I was fascinated to find that as a set or cohort of animals those calves that were born during the drought in 1976 fared badly all along, even years after that drought occurred. Of the 29 calves that were born that year, 14 died when they were less than a year old. Of the remaining 15, only 6 are alive today and 3 of these were born in December 1976, after the drought was over. It would seem that an adequate diet in its first year is crucial to the health and strength of an elephant for the rest of its life.

I found those trips to Amboseli in November and December very depressing. Each time I went out, there were more animals missing and presumed dead. The bad luck and difficult times I was experiencing in Nairobi seemed to follow me to Amboseli as well. In September it was decided against my own feelings of loyalty that Masaku and his brother Mwanga should retire. Masaku was not happy unless I was there, and Phyllis and Keith found him more and more exasperating to deal with. I agreed because I could see that the situation was bad for everyone. I contacted all the past and present sharers of the camp and we put together a large lump sum of money for Masaku's retirement. The final reckoning and parting were extremely distressing for me. Among other things, I felt that it was the end of an era in Amboseli and that the camp would never be the same again.

After Phyllis and Keith left, Marc Hauser, who was working on the vervet project, joined the camp. Marc is a lively, bright academic, who is also an excellent French cook (he is American but his father is French). Together we found a new worker, Peter Kumu Ngande, to take over Masaku's duties. Peter worked out very well and is still with us, but it is true that the camp is different.

The final blow for me in 1984 occurred on December 8. There was a ferocious wind storm in Amboseli that afternoon, and one of the large acacia trees in the middle of the camp snapped off about ten feet from the ground and pivoted just enough to fall directly on my Land-Rover and tent, crushing both beyond repair. Fortunately no one was hurt. I was in Nairobi; otherwise I almost certainly would have been killed. Peter was in the kitchen but only a few minutes before he had been in my tent zipping up the windows in anticipation of the rain to come. Marc, who had been out working, came back shortly afterward and experienced the full horror and shock of the scene when he drove around the corner and into the camp. He was very upset and immediately drove to Ol Tukai to try to get in touch with me. He managed to reach me by radio-phone that evening.

I very rarely cry, but I sat sobbing on my bed after talking to Marc. It had been difficult enough giving up my research permission, moving to Nairobi, and visiting the elephants for only short periods. Then retiring Masaku had been sad and distressing. Now it seemed that what was left of my life in Amboseli was destroyed. I had been very fond of that old Land-Rover. I had gently cared for it for nine years and knew all its foibles and idiosyncrasies. I had spent some of the happiest days of my life sitting in it watching elephants. It was very painful to realize that I would never again drive out of camp in it, and never sit in it again surrounded by peaceful, accepting elephants who had grown to know it so well.

Aside from the sentimental aspects, which had the most immediate impact, there was the financial side and the serious problem of having no transport in Amboseli. I still needed a tent and vehicle to carry on my work as coordinator of the project. I had very little money left in the AWF Amboseli Elephant Project account. Ironically, I had recently spent most of the remaining funds in the account to fix up the Land-Rover in the hopes that I could keep it going for another few years. It was 17 years old and it had needed a considerable amount of work. I had it panel-beaten and completely resprayed; I had new cushions, seat covers, window channels, springs, shock absorbers, an exhaust system, a battery, and new tires fitted; I had the electrical system overhauled and many smaller things repaired or replaced. I had spent $1,700. I could not claim anything back on insurance because, being such an old car, it was only third-party insured, and even if it had been comprehensively insured the company would not have paid because a tree falling is considered "an act of God."

The tent was ripped in several places and after nine years in Amboseli the canvas was too fragile to sew. The protective thatch roof was smashed and had to be dismantled. Perhaps the saddest thing of all was that we had to cut down the two other big trees in the center of the camp. These two trees were nearly dead anyway. It was the living tree with leaves on it that fell, probably because it offered more resistance to the wind.

I could not face the Land-Rover and the trees, and spent Christmas somewhere else for the first time in many years. I flew down to Amboseli on December 28 and stayed in the study tent that Phyllis had donated to the project when she left. I was able to drive the small Suzuki pickup that she had used, which was temporarily being stored until Sandy came to start her work. The camp looked stark and bare and bright to me without the trees, and the wrecked Land-Rover and all the other debris—broken glass, smashed wood, rotting thatch, and thorns and branches from the felled trees—depressed me. I did not enjoy the stay but it was important to get out and census the elephants before the end of the year.

Joyce and I also had to "do" the vegetation plots. Neither of us is an ecologist and we dreaded the slow and tedious work involved in monitoring vegetation. Keith had set out these plots to determine the availability of food for the elephants during the two rainy and two dry seasons in each year. It was data we all needed and so we agreed to keep up the monitoring after he left. This was our first time to do it. To start with we had to find the 12 plots, which were scattered in the various habitats throughout the park.

Each plot consisted of a transect line delineated by two stones 100 meters apart. We had to find the first stone by a combination of memory and Keith's map and then take a compass reading to find the second stone. A few of them were easy; most were extremely difficult, with rocks buried in tall grass or hidden among bushes. Once we found the rocks we ran a string out from a special winder that Keith had made. The string was marked off at two-meter intervals, giving us 50 points along the line at which we had to take measurements. Keith had designed a metal frame with ten vertical rods. We had to measure how many of these rods hit a piece of grass or herb and then measure the height of the vegetation. That made over 500 individual measurements per plot.

We set out on December 29 to start the plots. We did two the first day, two more the next day; and on the third day, New Year's Eve, the final disaster of 1984 occurred. This one, however, was amusing. Joyce and I left early, determined to get six plots done. We ound the first with no trouble and quickly finished it and then drove to one of the swamp-edge plots, where the bush was fairly thick. It was a dangerous place and Keith warned us to be careful there. It was a favorite spot of lions, buffaloes, and elephants. We found the first rock, made a compass reading, and set off in the direction of the second rock, whistling and clapping our hands to try to scare any unfriendly animals away. We did not come across any animals but we also could not find the other rock and wandered around, crawling under bushes, getting scratched by thorns, and cursing Keith. Eventually we found it, ran the string out, over, around, and under the bushes and began taking the measurements. We were three quarters of the way along the line when we heard the unmistakable flap of an elephant ear. Unfortunately, we could not see anything, since we were surrounded by bushes well over our heads. We went back to the vehicle and climbed on top of it. Coming along in our direction was a cow/calf group. We drove the car to the end of the line and very quickly tried to finish the transect. As the group got closer I stood on top of the car to warn Joyce when to run. Just as she started to rewind the string, another elephant came along behind her and I shouted for her to come quickly. She had to abandon the winder and string and sprint for the car.

We both stood on top of the car watching the cows and calves approach the beginning of the transect. We could see that it was the DB family, led by Delia and Deborah. We hoped that they would just pass by or over the string, but no such luck. Daniel, Deborah's seven-year-old son, found it first and stopped short in horror. (Elephants notoriously "dislike" anything new and foreign in their home range; much to the annoyance of park wardens they almost always knock down signs unless they are surrounded by a huge barrier of rocks.) After the initial surprise Daniel approached the string, grabbed it in his trunk, lifted it, and pulled on it and then rapidly began winding it around his trunk like spaghetti around a fork. He turned around and backed into the string and wound some more around a back foot. We could not help but laugh. By now the others had joined him and also began pulling on the string. Even the big females took part in the game. Suddenly at our end of the line, Joyce spotted the little black marker dots flashing past, followed by the winder rapidly bouncing along the ground and over bushes toward the elephants. The whole group of elephants, with string wound around limbs, trunks, and tusks, went off behind some other bushes and stayed there for another ten minutes, apparently playing with the string and winder. Eventually they moved on to the swamp.

Joyce and I cautiously got down from the car and began assessing the damage. We found at least ten different sections of broken-off string, some caught up in branches, others simply in knots. We collected these but knew it was hopeless to try to piece them together again. What we really wanted to retrieve was the winder, which was specially designed for the work. Keith had taken a great deal of care and time making it from all sorts of scraps and bits and pieces. We searched for nearly an hour and could not find even a piece of it. We dreaded having to write and tell Keith. Nevertheless we still thought the whole thing was extremely funny and kept laughing every time we thought of Daniel winding the string around his trunk or the winder bouncing and skipping by us. The next day we did not think it was quite as humorous when we went out with a new string and a primitive winder that consisted of merely a stick. It made the whole process even more tedious, but we—and later Joyce and her new assistant, Nellie Sakawa—managed to finish the plots in the first few days of the New Year.

In the meantime, 1984 did not come to an end without the Ts experiencing yet another disaster. Titus, Masaku's 22-year-old son, who was also working as one of our research assistants, reported to me in early December that Tilly's '83 calf had a very stiff right hind leg. When I saw the calf on December 28 his leg was badly swollen and he could not put any weight on it. It bulged out at the joint and looked as if it was dislocated or broken. It was pitiful watching him trying to move along on three legs, clearly in pain. We thought it was just a matter of time before he died. I suspect he received the injury by falling into the big garbage pit, so it looked as if yet another T was lost to garbage-related causes.

The Ts had suffered so many losses in 1984 that I almost felt that they were jinxed. I hoped that 1985 would bring them better luck. There had been only one birth among the Ts in 1984: Tuskless had a female calf on March 20. In total there were seven deaths—Tania, her C'83, Teresia, Tia, her C'83, Tio, and Tad—and two young males, Toby and Tolstoy, became independent. Reduced by eight, the three remaining families numbered 21 animals at the end of 1984:

TA	Tuskless (62)	♀
	C'84	♀
	Tuo (80)	♂
	Tonie (67)	♀
	C'82	♂
	Tilly (70)	♀
	C'83	♂
	Tom (79) [ex-TB]	♂
TC	Slit Ear (36)	♀
	C'83	♀
	Slo (80)	♀
	Tabitha (76)	♀
	Tara (67)	♀
	C'82	♂

	Tess (63)	♀
	C'82	♀
TD	Tallulah (63)	♀
	C'83	♀
	Tao (80)	♀
	Theodora (67)	♀
	C'82	♂

Chapter XI

Future Generations

1985–1986 and Beyond

Slit Ear lifted her head, flapped her ears, then folded them horizontally and lunged at Gladys, jabbing her in the backside. Gladys sprinted away, uttering a groaning bellow. Tara came up beside Slit Ear, flapped her ears, and rumbled, but Slit Ear ignored her. They both resumed feeding.

The TCs were in an aggregation of about 45 elephants in Longinye. The other T families were not present. Slit Ear's family was small now, consisting of only eight animals: Slit Ear, her C'83, Slo, Tabitha, Tara, her C'82, Tess, and her C'82. The families they were with were all larger: The GBs, of which Gladys was a member, contained 17 elephants and they formed a close bond group with another family there, the IBs, with 13 animals. The IBs' matriarch, Isis, was big, close to the same age as Slit Ear. And the fourth family there, the LBs, was led by another big female, Leticia. Still, Slit Ear asserted her dominance over them all, and while she was feeding and moving through the aggregation, the other cows and calves gave way to her, discreetly moving out of her path with a wary look over the shoulder.

285

The four families remained together, feeding throughout the morning and stopping to rest and dust at midday. In the afternoon the TCs gradually drifted away from the aggregation and headed west toward the Ol Tukai Orok woodlands. They moved slowly, feeding as they went, but when they reached an open short-grass plain they formed a line with Slit Ear in the lead and Tess at the rear and walked with a deliberate, long stride. They crossed the plain and then the main road and then a nearly bare pan and entered the woodland.

Here they met up with the TAs and Tom but there was no exchange of friendly greetings. Instead, Slit Ear folded her ears at Tuskless and Tonie, walking rapidly toward them. They moved smartly away rumbling, and one of them gave a short trumpet of alarm. Slit Ear approached in an aggressive posture again and once more Tuskless and Tonie hurried away. A few minutes later Tara suddenly swung around toward Tom and tossed her trunk at him. Tom gave a groan-scream and ran away. She did it again and he bellowed, although she never touched him.

The two families did not stay together for long. The TAs circled back the way they had come and Slit Ear led her family on through the woodlands and out to the south into the lava, toward the mountain. She did not stop and call or listen for other elephants. The eight animals spent the night feeding and resting. Of the adults Tara stayed closest to Slit Ear. Tess kept her distance, especially when they were feeding.

In the morning they moved back toward Ol Tukai Orok, stopping to rest in a patch of *consimilis* on the plain to the south of the woodland. While napping in the morning sun, Tara suddenly lifted her head and listened. She rumbled very low and with ears out listened again. This time she emitted a louder, more throaty rumble. She and Tess began to secrete from the temporal glands and turned toward the south. Slit Ear did not respond.

Two minutes later Tara and Tess rumbled, listened and rumbled again. Another three minutes passed and a small group of five elephants appeared on the plain to the south. It was Tallulah and her two calves and Theo and her calf. They walked steadily, coming straight for the TCs. The four young adult females, plus nine-year-old Tabitha, came together and greeted each other with raised heads, flapping ears, and a few rumbles in a friendly but subdued ceremony. Slit Ear did not participate at all, although her five-year-old daughter, Slo, ran over to greet Tao, her age-mate. The two calves entwined their trunks and placed them in each other's mouths. Tess's three-year-old daughter cautiously approached these older calves and was fondled and touched and inspected by both of them. Tara's and Theo's male calves almost immediately began a sparring match, neither of them having had another male of the same age to play with for several days.

Slit Ear did not display any aggression toward Tallulah and Theo, as she had with Tuskless and Tonie, but she also showed little interest in them. She continued to rest while the others were socializing. Eventually the five TDs settled down to nap as well, but unlike in former times they did not stand bunched together with the TCs. Instead, they stood off to the side, forming a distinct group of their own. About 15 minutes later Slit Ear woke up, stretched, urinated and defecated, gave a perfunctory "let's go" rumble, and set off toward the woodlands. All the others followed.

In the months after the deaths of Teresia, Tia, and Tania in 1984 I watched with some dismay the profound effect those losses were having on the social fabric of the T bond group. Without Teresia, the TCs and TDs disintegrated and fragmented. Tallulah and Theo were rarely with Slit Ear and they were sometimes not even with each other. Slit Ear's own family broke down much of the time, with Tia going off on her own until her death in October. Afterward I often found the TCs split up into two or even three groups.

Some of this behavior can be attributed to the drought, during which many families splintered into mother-calf subunits, but the TCs and TDs had been through an equally bad drought in 1976 and had stayed together. Even after average rains fell in 1985 the bonds between the two families did not seem as strong as in the past. Each of the families itself coalesced again but I still found the two families separated on several occasions, whereas before it was a rare event to find them apart. Intuitively, I felt that the bond that had held the TCs and TDs together so consistently was the relationship between Slit Ear and Teresia. For whatever reason the allegiance between them was exceptionally strong. Without the attraction that Slit Ear and Teresia had for each other a realignment and shift of loyalties occurred.

Far more striking was the change in the relationship between the TAs and the other two T families. I think there must have been a strong bond between Slit Ear and Torn Ear, and a barely weaker one between Teresia and Torn Ear. The data on association patterns before 1977 suggest that there was. However, once Torn Ear died, the affiliation between the families diminished. Possibly there were some residual bonds between Tania and the other older generation of females, but then she died too, leaving only the young TA females. Slit Ear apparently felt no attraction to them and actually appeared to discourage close contact. It was on January 2, 1986, that I saw her acting antagonistically toward Tuskless and Tonie and watched Tara tossing her trunk at Tom. Thus, by the beginning of 1986 I could barely justify calling them members of the same bond group. In fact if I had started my study in 1986 I would never have guessed that the TAs had a special relationship with the TCs and TDs. They spent little time together, they rarely greeted one another, and Slit Ear was antagonistic toward them, hardly the right atmosphere for bonding. Still, the history of their relationship was there in the records and I could not dismiss the possibility of further changes and realignments occurring.

I cannot help but wonder what the Ts' history was before my time and what events and factors brought about the relationships between Slit Ear and Teresia and Slit Ear and Torn Ear. Was something forged in Slit Ear's childhood? Was Teresia her older sister or her aunt? Did Torn Ear care for her when she was a small calf? Did the bonds persist because of a consistency in behavior, a knowledge that they would always help each other? And what about the simple factor of time? All evidence indicates that Slit Ear had spent most of her life in the company of Teresia and probably a great deal of time with Torn Ear as well. Could shared experiences over the years maintain and cement the bonds? Do elephants really never forget?

I cannot answer those questions. I can only watch Slit Ear now and try to interpret her behavior. Although she does have her daughters and her family around her, she seems different to me. I have not done a quantitative study of her social activities but my impression is that she has displayed more aggressive behavior and far less affiliative behavior in the time since Teresia died.

I am curious, though, to see what will happen when Slit Ear dies. She is the last of her generation. When Teresia died, Slit Ear was 48 years old. As I write this now, she is 51. All the remaining females in the three T families are less than half her age. Tuskless is the oldest, but only 25; Tess and Tallulah are 24; Tara, Theo, and Tonie are 20; Tilly is 17; and Tabitha is 11. They are the new generation. The bonds they have formed and that continue to develop will determine the social cohesiveness within and between the families. There appear to be fairly strong ties among these younger females. Perhaps when Slit Ear dies they will reestablish closer affiliations and spend more time together. For the time being Tuskless leads her little group, Slit Ear is the dominant figure in her family, and Tallulah has definitely become the matriarch of her family.

Although I would have liked to have followed the lives of the Ts more closely in the two years following the deaths of 1984, I was by necessity writing in Nairobi for much of 1985 and 1986 and did not see the Ts often. In general it appears to have been a relatively uneventful period for them, a respite from the horrors of 1984 and a time of recovery.

When I did see them, each time I was surprised to see that Tilly's '83 calf, who had broken his leg, was still alive. It was sad to see him hobbling along. His leg was withered and there was a huge swollen knob at the top joint. Yet he was feeding well and he was fat. Tilly slowed her own pace or always waited for him to catch up. One day I was touched to see him playing with a calf from the SA family. They pushed and butted one another and I am sure he was at a disadvantage but he was sparring with spirit and verve.

In September 1985 I saw Teddy with a hugely swollen penis and sheath. My colleagues Joyce and Sandy had noted it a week earlier and they thought he might have been bitten by a snake. Miraculously, after several weeks he recovered. The other T "boys" all seemed to be doing well: Right Fang, Tim, and Toby spent much of their time in and around the camp. Unfortu-

nately Toby was addicted to human foods and prowled around
Ol Tukai at night, knocking over garbage cans and walking
right up to the verandas and windows of the bandas. Even we
had some trouble with him when he tried to knock our kitchen
down several times toward the end of the dry season of 1986. I
had a small electric fence put up around the kitchen and that
seemed to discourage him. Teresia's son, Tolstoy, went from a
family member to independence very rapidly after his mother
died. I often saw him on his own, with other young males, or on
the periphery of family groups. As he was growing older his
resemblance to Teresia was striking. With his long, almost
straight tusks, I think he will be a very handsome male when he
reaches his prime.

During 1985, 65 females were recorded in estrus, and no
doubt others were missed, which led to a mini baby boom in
1987. Among those who were seen in estrus in 1985 was Tonie.
She was in consort with and seen mated by Bad Bull in July. In
May 1987 another calf almost certainly fathered by Bad Bull
entered the T bond group. Once more the relationships were
labyrinthine because this calf was the half brother or sister to
Slo and possibly others in the TCs and TDs on the male side
and was the cousin, niece, and sister or brother to others on the
female side. In April 1986 ten-year-old Tabitha came into
estrus and was mated by Pablo. In October 1986 I noted that
Tess was looking fat and was probably pregnant. Thus it seems
that many new little Ts are coming into the world. I only hope
they are able to live out their lives relatively undisturbed in an
atmosphere where the full range of their behavioral and social
repertoire can be played out. I have my doubts and I worry a
great deal about their future.

Amboseli's elephants are apparently safe for the next two or
three years, but beyond that I honestly cannot say what will
occur. Events affecting Africa's elephants on a larger scale
across the continent will eventually impinge on the Amboseli
elephants. Ivory poaching, loss of habitat, and culling schemes
loom on the horizon. All these factors are inexorably working
against the future existence of a relatively natural elephant
population such as Amboseli's.

The most acute and immediate threat to the future of
elephants in Africa is the ivory trade. Ivory is no doubt a

beautiful substance. It has an incandescence and warmth unlike any other material used for ornaments or sculpture. But I always feel that people forget that ivory is the tusk of an elephant. The word *ivory* disassociates it in our minds from the idea of an elephant. One tends to lump it with jade, teak, ebony, amber, even gold and silver, but there is a major difference: The other materials did not come from an animal; an ivory tusk is a modified incisor tooth. When one holds a beautiful ivory bracelet or delicate carving in one's hand, it takes a certain leap of understanding to realize that that piece of ivory came from an elephant who once walked around using its tusk for feeding, digging, poking, playing and fighting, and furthermore that the elephant had to be dead in order for that piece of ivory to be sitting in one's hand. I would venture to guess that only a tiny percentage of people who buy ivory ever make that leap. Even if they did they would probably buy ivory anyway in the all-too-common belief that there are plenty of elephants in the world, possibly too many, and in any case they are here for the use of man.

Trade in ivory has been going on for centuries; as a consequence, elephant populations have fluctuated and in many instances have been exterminated. Approximately a thousand years ago North Africa's elephants were wiped out for their ivory. (It was from this population that Hannibal obtained the elephants that he used to cross the Alps in c. 219 B.C.) In South Africa there were only a few small pockets of elephants remaining by the nineteenth century after a period of intensive and unchecked ivory hunting. And in West Africa most elephant populations had been eliminated by the turn of this century. A few remnants, many of only 20 to 30 animals, were left dotted about. In eastern Africa elephant populations were drastically reduced in the nineteenth century because of the conjunction of the ivory trade and slavery. The two trades fed on each other as the ivory traders captured men and women to carry tusks to the coast for shipment and then sold the people into slavery. At the height of the ivory trade, approximately 1,000 tons of ivory were being exported to Europe every year.

In the early twentieth century some colonial administrators became alarmed by the rapid depletion of elephant populations. Laws were instituted on the hunting of wildlife and reserves were set aside where animals were protected. As a result elephant populations began to recover and increase in eastern, central, and southern Africa. However, rapid human population growth in West Africa caused a further decline in that region's elephants. The ivory trade continued, with the major importing activity shifting from Europe to eastern Asia. For half a century, the volume of ivory traded stayed at a relatively low level, with prices remaining stable until the 1960s. However, in the late sixties, the price of ivory, which had long been approximately $2.45 per pound (or at that time about one British pound per pound of ivory), began to increase. For a variety of reasons which can only be guessed at, ivory once again became a popular form of hard currency. By 1973 the price had risen to nearly $14 per pound and by 1978 it had reached $35 per pound. Even taking inflation into account, the real value of ivory was six times higher in 1978 than in 1960. Some suggested factors behind the upsurge in the trade were: the increased availability of elephants, which were once again numerous in areas where they had formerly been nearly wiped out; the unstable political situation in some countries that had recently gained independence; the increasing human populations, with subsequent conflict between elephants and man; and the general chaos in the money markets caused by the skyrocketing prices of oil, suddenly making ivory a valuable and stable commodity.

The demand for ivory has increased even further in the 1980s and today the price is over $50 per pound. The total trade in raw ivory per year is presently around 825 tons. Big money is involved. The trade in raw ivory is worth at least $50 million a year and the trade in carved ivory is worth $500 million a year. Japan is the biggest importer (523 tons in 1983), with virtually all the ivory marketed internally. It is used to make signature seals, which the Japanese prefer to writing out their names, jewelry, decorative carvings, and netsukes, which are worn on kimonos. A smaller proportion goes to other Asian countries and to Europe, and about 10 percent remains in Africa, where ivory carving is a rapidly growing industry in several countries.

These facts and figures on ivory, even in my mind, tend to distance one from the elephant, but ivory statistics do not represent the harvesting of so many trees or the mining of a quantity of ore; they represent a definite number of dead elephants. Of course, some elephants die of natural causes and their tusks are collected to be sold by the wildlife departments or central governments involved. However, natural mortality at this point cannot possibly feed the current demand for ivory. To meet the demand for 825 tons of ivory per year a minimum of 70,000 elephants have to die. (This figure is based on the recently recorded average tusk weight of approximately 13 pounds and an average of 1.8 tusks per elephant, which accounts for one-tusked and broken-tusked animals.) And that is only the elephants with tusks; thousands of calves also die because of the loss of their mothers, bringing the estimated total up to at least 80,000 elephants each year. Africa's elephants, currently estimated at 800,000, cannot possibly sustain that level of offtake.

The total number of elephants in Africa and the effect that the ivory trade has been having on them have been issues of hot debate among conservationists and people with an interest in the ivory trade in recent years. From 1981 until only recently, the ivory lobby had convinced even many conservationists that the trade in ivory was having no significant effect on the overall number of elephants in Africa. Today it is clear that this view did a great deal of damage to attempts to conserve elephants. The indisputable evidence coming in from across Africa is so disturbing and so overwhelming that there is no longer doubt in anyone's mind that Africa's elephants are being killed at an unsustainable rate. The most recent figures reveal that they are declining at 5–10 percent per year depending on the region.

Iain Douglas-Hamilton has been involved in surveying the status of elephants throughout Africa since 1976 for the IUCN (International Union for the Conservation of Nature and Natural Resources). From 1976 to 1982 he was chairman of the IUCN/Species Survival Commission's African Elephant Group and later he continued his surveys as a consultant. He published a summary of the African elephant's status in the January 1987 issue of the conservation journal *Oryx*. In all but a few countries, whether in parks and reserves or not, elephant numbers are declining. The only notable exceptions are South Africa, Zimbabwe, Botswana, and Malawi, where elephant numbers are either increasing or remaining stable.

Some areas are experiencing far more drastic declines than others. There is a crescent of intensive elephant poaching which stretches from the Central African Republic, through Chad, Zaire, southern Sudan, and Somalia. To give a few examples:

Central African Republic

Iain carried out aerial surveys of the Bamingui-Bangoran and Manovo-Gounda national parks in 1985 and discovered an 80 percent decline since the previous count in 1981. In 1976 there were an estimated 80,000 to 100,000 elephants in the country. The 1985 estimate was 10,000 to 15,000.

Chad

Before the outbreak of civil war in Chad, the overall elephant population was estimated at 15,000. The latest estimates suggest that fewer than 2,000 remain.

Sudan

Gangs of 60 and more poachers with automatic weapons have all but wiped out southern Sudan's elephant populations.

Somalia

Aerial surveys in Somalia have revealed far more carcasses of elephants than live elephants and one observer believes that there may be only a few elephants left there now.

In other areas of Africa large numbers of elephants were exterminated in the previous decade and in some cases the slaughter continues. Again just a few examples:

Uganda

In the 1970s, during Amin's reign and the chaos of civil war, 80–90 percent of Uganda's elephants were killed. In Murchison Falls (South) National Park, where accurate aerial counts have been made for many years, there were 9,970 elephants in 1973; nine years later, in 1982, there were only 25 left.

Kenya

In 1974 Kenya's elephants were estimated at 167,000. By the end of the seventies 100,000 were gone. Today there are estimated to be only 35,000–40,000 and recent reports indicate that renewed intensive poaching is under way again.

Tanzania

Ivory poaching in Tanzania only began to be severe in the late seventies, but it has been virtually out of control in the eighties. In the Selous Game Reserve in southern Tanzania Iain counted 110,000 elephants in 1976, making it one of the last strongholds for elephants in Africa. In October 1986 he did another count and found only 55,000 elephants there.

It is not the numbers alone that are depressing; equally disturbing is what has happened to the populations that remain. Poachers systematically take the animals with the largest tusks, so the first to go are the males and the older matriarchs (the Japanese carvers prefer large tusks and actually pay more money per pound for large tusks than for medium and small tusks). When the older elephants are finished they start on the medium-sized ones and eventually have to do with young elephants with very small tusks. Not surprisingly, the average tusk weight exported from Africa has dropped significantly in the last few years. According to the Wildlife Trade Monitoring Unit of the IUCN, the mean tusk weight of ivory imported into Japan dropped from 35.8 pounds in 1979 to 21.3 pounds in 1982. In some of the most heavily hunted populations it appears that only young animals remain. Iain recently recorded mean tusk weights of ivory confiscated from poachers in Uganda and the Central African Republic at 5.9 pounds for Uganda and 6.6 pounds for the CAR. Elephants with tusks in that weight range would be no older than seven or eight years. Overall, the average tusk weight of recent export records is about 13 pounds, which would be an adult female or a teenage male. In fact, it is estimated that the majority of the tusks presently in the trade are from females in their reproductive years.

What these data on mean tusk weights reveal is that most

elephant populations are being seriously overexploited. To fulfill the demand for the 800-plus tons that are needed for the trade, the suppliers have to kill more elephants each year. Where ten years ago possibly 40,000 elephants with an average tusk weight of over 20 pounds could fill that need, today, with much younger populations, as many as 70,000 elephants must be slaughtered. Each year more elephants with smaller tusks will be needed to satisfy the trade. The average age of the populations will get younger and younger and inevitably less productive.

My work in Amboseli, as well as the work of others on social organization, mating behavior, calf development, and population dynamics, indicates that a population consisting of only young animals under constant stress cannot possibly reproduce and raise calves at even an average rate. The most productive and successful elephant mothers are those in their middle years, between 25 and 45 years old. Even under good conditions the calves of teenage females have a higher mortality rate than those of more experienced females. Under heavy poaching pressure with leaderless, fragmented families, the calf mortality rate is bound to increase. I would further predict that the conception rate would decrease from stress and in the absence of older musth males. In a relatively undisturbed population elephants can increase at 3–7 percent per year, but it would be unlikely that a young, harassed population could breed at anywhere near this rate.

The situation is depressing but it may not be irreversible. There are people working very hard to halt the downward trend. Iain Douglas-Hamilton continues to labor unflaggingly to try to make the public aware of what is happening to elephants. Others have joined the battle. The chairmanship of the combined IUCN/SSC African Elephant and Rhino Specialist Group (AERSG) passed on to Jonah Western in 1982, and he initiated some valuable studies, particularly those dealing with the ivory trade and with modeling the various harvesting methods for ivory. He also instigated a study and survey of the forest elephant in Central Africa and helped influence Sudan to ban the sale of ivory. In 1985 David Cumming of Zimbabwe's Department of National Parks and Wildlife Management became chairman of AERSG and he too has worked extremely hard to try to halt the decline of both rhinos and elephants.

Not only conservationists are concerned. African governments, realizing that they were losing revenue and a natural resource to poachers and illegal traders, approached CITES (Convention on International Trade in Endangered Species of Wild Fauna and Flora) officials and requested an ivory quota system, which has been in operation since January 1986. A special ivory unit of the CITES secretariat in Lausanne, Switzerland, is now able to monitor the international trade in registered tusks. All legally traded ivory must be marked and accompanied by CITES documentation and cleared through their office. Theoretically, since the exporting nations of Africa and most of the major importing countries have signed CITES, it should be possible to keep track of all the tusks on the market, from source through the various intermediaries to their final destination. In practice there are still many loopholes, and the decline in elephants has continued unabated since the quota system was instituted.

In an attempt to approach the problem from the importing end, Esmond Bradley Martin, an American geographer, has been carrying out a project that has opened up a dialogue between conservationists and ivory dealers. Martin has visited all the countries that import ivory in any substantial quantity. He has found in many cases that the ivory traders and carvers themselves are aware that they may be out of business in a few years if some changes are not made. He has had a fairly positive response in Japan, where the Ivory Division of the Japan General Merchandise Importers' Association contributed funds toward the setting up of the ivory unit in Lausanne.

Basically what the ivory traders want is a sustained yield of tusks coming from Africa each year. Ironically, to achieve this goal by far the best solution would be, for most areas of Africa, not to kill any elephants at all. This highly interesting conclusion has come from computer simulation modeling carried out by Tom Pilgram, a statistician, and Jonah Western. Taking all the information available on elephant population dynamics—numbers, age structures, reproduction and mortality rates—and on growth rates of tusks, Tom and Jonah came up with various "harvesting" models. They then compared their different models with the data on tusks that have passed through the ivory trade in the last few years. The one that fits the present trade figures is one that takes more and more elephants each year from a younger and younger population in order to produce a constant total weight of ivory. In this model elephant populations decline and crash to local extinction. Tom and Jonah predict that this simulated result could become reality for at least half of Africa's elephants within a decade.

In another model, the same number of elephants are killed each year and eventually the population stabilizes but at a younger age, which results in the total ivory harvest declining and mean tusk weight remaining low. The model that would eventually produce the most ivory and the biggest tusks would be one in which no elephants were hunted for the foreseeable future, populations were allowed to recover, and, most important, could reach older ages and die of natural causes. This somewhat unexpected conclusion comes about because tusks continue to grow throughout an elephant's life, and in males the rate of growth speeds up in the last decade of life.

It is unrealistic to expect a moratorium on the killing of elephants or even a moratorium on trade. With $500 million involved someone will always find a way to get ivory out of Africa. Attempting to control the ivory trade is truly a huge task, given the political and social climate in Africa. The political situation in many of these countries is so precarious that almost any control is out of the question. Elephants in countries engaged in civil war are particularly vulnerable.

Iain Douglas-Hamilton has completed a valuable investigation of the relationship between the increase in sales of automatic weapons to African countries and poaching for ivory. In his recent *Oryx* article he reports that according to the U.S. Arms Control and Disarmament Agency the value of arms imports to Africa increased (in real terms, taking inflation into account) from $500 million in 1971 to $4,500 million in 1980. In the same period armed forces in the East African region increased from 141,000 to 441,000. In general, where there has been increased importation of arms there has been more elephant poaching. In some cases the armies themselves have been the main poachers, as was the case with, for example, Amin's troops in Uganda. In other cases civil wars are being partially financed by ivory poaching. In yet other countries the mere availability of automatic weapons, which have easily found their way into private hands, has changed the equation. In many areas where government antipoaching forces have previously been able to keep poaching in check in national parks and reserves, they are now powerless to oppose well-organized and far better armed gangs.

In countries that are not at war, the major problem is corruption of officials. It is a simple matter for a customs or wildlife officer to fill in and sign the CITES forms making poached ivory legal. If that is not possible illegal ivory can be sent out through the central African country of Burundi, which is not a signatory of CITES. Burundi exports most of the ivory coming out of eastern Africa although it has no elephant population itself. The tusks come from poached elephants in Zambia, Tanzania, Zaire, Uganda, Sudan, Kenya, and even farther afield. The tusks are sent on to the United Arab Emirates, which is the only country in the world that has ever signed and then withdrawn from CITES. With Burundi and the UAE free of any controls, corrupt officials in the African countries with elephant populations will always have a market for poached tusks.

Trying to tackle widespread corruption in African governments is too formidable a task for conservationists, but the corruption within the wildlife departments might be reduced if more financial and moral support was given to national parks and reserves. While many elephant populations still exist outside their boundaries, in the end it is in the so-called protected areas that conservation efforts in Africa will have to be concentrated. But with a few exceptions the wildlife departments operate on minimal budgets, and wardens and rangers are pitifully underpaid and underequipped. When the price of one pair of tusks is equivalent to an officer's yearly salary, the temptation to poach is overwhelming. We in the developed countries could help Africa's conservation areas by contributing far more to their maintenance. It would be a crime to future generations to stand by and let parks and reserves continue to deteriorate. If war and anarchy, greed and corruption cannot be avoided, lack of funds is remediable. There are many, many African people who care a great deal about their wildlife and some have lost their lives trying to protect it; but their countries are poor and revenue that may have been generated by wildlife often has to be used for health or education, or, unfortunately, arms. In any case, the burden of wilderness protection in Africa should not be placed exclusively on the backs of the African people—these areas and their wildlife are the world's heritage.

I, of course, have to ask myself what all this means for Amboseli and the elephants whose lives I care so deeply about. Sometimes I try not to think about it because at the moment in Amboseli there are no gangs of 60 men carrying AK 47s and mowing down every elephant with tusks. But I also know it could happen, given the right set of circumstances. Kenya is a stable, peaceful country; but if there was serious political unrest or civil war in Kenya or neighboring Tanzania, Amboseli's elephants would represent a quick source of revenue for buying arms. Until ivory poaching becomes too costly to the hunter because of danger in coming up against well-armed and well-trained rangers, or because the selling of illegal ivory becomes difficult, or because the price of ivory goes down through lowered demand, Amboseli's elephants will never be safe from the threat of poaching.

What I feel is needed now is a worldwide campaign to reduce the demand for ivory. I would ask people not to buy ivory until a time comes when one can be sure that the tusk is either from an elephant killed legally or from one who died of natural causes. I know that the stores carrying ivory in the United States claim that they buy only ivory that was procured legally. Yes, the ivory may have entered the country legally, with all the correct papers, but the overwhelming chances are that it left Africa under suspect arrangements. It is currently estimated that a minimum of 80 percent of the ivory traded today is from poached elephants.

While ivory poaching is the overwhelming threat to Africa's elephants, two other major threats also have to be considered. Not as immediately acute as poaching but in the long term just as perilous to elephants is the problem of population growth and the concomitant expansion of people into wilderness areas. Elephant ranges are shrinking every day while Africa's human population is growing prodigiously. In Kenya, which has the highest recorded birthrate in the world, at close to 4 percent per year, the population, last censused at 15 million in 1979, will double by the year 2000. Where will all those people go? Less than one third of Kenya's land is arable. The rest is semidesert, uninhabitable except by only the toughest of nomadic pastoralists. Nearly 7 percent of Kenya is gazetted as national parks and reserves, which is high. However, in some areas people live right up against the boundaries of parks and are petitioning their members of parliament to degazette the parks in order to let them graze their cattle or cultivate crops inside. It may just be a matter of time before the government has to give in to their pleas. On the other hand, tourism is Kenya's second-largest foreign-exchange earning industry, exceeded only by coffee. It is in the interest of the country as a whole to maintain its conservation areas, but hungry, landless, unhappy people cannot be ignored. The problem is one that only the Kenyans can solve.

Changing patterns of life-style and land use among people are also affecting wilderness areas. Around Amboseli, because of population growth, the Maasai are gradually shifting to a mixed agricultural and pastoral economy. Previously the Maasai could manage a subsistence existence based solely on traditional pastoralism if they had approximately ten cows per person. Now there are more people and the land they range over can sustain only four cows per person. The Maasai have little option but to enter the money economy to supplement their diet with agricultural products. I am not trying to imply that the Maasai should have stayed traditional pastoralists or that the change is necessarily a bad one, but there is no doubt that the change is accompanied by a different attitude toward wildlife. Wild animals, particularly elephants, may be able to coexist with pastoralism, even a modern form of pastoralism, but they cannot coexist with agriculture.

Aid organizations are continually trying to get pastoralists to become agriculturalists or fishermen or whatever, to settle them in villages and towns and stop their nomadic way of life. There have been more failures than successes in these attempts—witness the spreading desertification of the Sahel, attributed by many authorities to a generation of ill-advised water-development schemes. Yet these organizations continue to try. Many are run by missionaries who, along with wanting to help primitive, pagan peoples, also want to convert them and have them attend church. A nomadic way of life is not conducive to churchgoing, but encouraging people to settle in a village or town is.

Presently in the Amboseli area a missionary group is helping the Maasai with an irrigation project at Namalog swamp, just to the east of the border of the Amboseli National Park. The Maasai, as well as people from other tribes, had already settled and farmed at Kimana swamp farther east, but Namalog had remained free of agriculture until recently. Now there are houses, fences, and irrigation channels. (There is also a church.) How long the Maasai will be able to grow crops at Namalog is open to question, for the soil may not be suitable and there is a limited amount of water. In the meantime the new agricultur- alists are crying out against the wild animals that try and often succeed in sneaking into the fields to eat the crops. If this irrigation scheme is seen as a success, then it is only a matter of time before the people in the area start to look longingly at the two big swamps—Longinye and Enkongo Narok—within the park.

However, it would not appear to be in the national interest to turn Amboseli over to agriculture. Amboseli has the highest gate receipts of any park in the country. Each acre of that park is worth far more each year to Kenya as wilderness than it would be under agriculture or ranching. Nevertheless, at least some individuals in the government are aware that local people living on the periphery of national parks and reserves have to benefit from those areas in order to tolerate and encourage wildlife. Without their cooperation a small park like Amboseli could not survive.

Since the 1950s, schemes to help the local Maasai realize the potential benefits from wildlife have been tried, some with more success than others. From early on the wildlife authorities have tried to provide water for the Maasai in the peripheral areas in order to leave the central viewing area free of cattle. When the status of Amboseli changed from a reserve to a national park, a water pipeline was built for the Maasai and other promises of revenue and assistance were made. Most of the benefits proved to be short-lived or never forthcoming. The pipeline was of inadequate design for the number of cattle in the area, and then it was shut down completely because of lack of funds to buy fuel for the pump. At the same time, the boreholes that the Maasai had been using for many years were no longer maintained and the Maasai found themselves with no water except for that in the swamps. New conflicts arose, with the Maasai justifiably angry because the government did not fulfill its promises.

The Maasai also lost important revenue in those years. Before Kenya declared a nationwide sport-hunting ban in 1977, the Maasai received hunting fees from people who wanted to shoot on their land. These fees were considerable; in the final year before the ban came into effect the group ranches surrounding Amboseli received 1.6 million Kenya shillings (about $225,000) from hunting licenses, mostly for buffaloes, antelopes, zebras, and birds. Although I never want to shoot any animal myself, I am not against sport hunting because it brings revenue to the people who have to live with wild animals, and taking a few trophy animals each year apparently has little detrimental effect on wildlife populations.

The Amboseli Maasai also received benefits in the form of grazing compensation based on the numbers of animals migrating onto their ranches in the wet season. These payments lasted for five years after the park became operational in 1977 and then were suspended. A proportion of the revenue from the lodges goes to the Kajiado County Council, which is then supposed to put the funds back into the area in the form of cattle dips, schools, and other development projects. Unfortunately, Kajiado is 100 miles away and the councilors have more pressing needs nearer home. Very little appears to get back to Amboseli.

As I write this, relations between the Maasai and the park could be a great deal better. Work on an improved pipeline has not yet begun. There is no longer any grazing compensation or revenue from hunting. The only direct benefits the ranchers receive are from the public campsite, from special camps used by luxury tented safari companies, and from the sale of firewood to the lodges and camps. A lot more could be done. The Maasai could be helped to be successful livestock ranchers by combining their traditional knowledge of semiarid environments with some modern production expertise, enabling them to participate more fully in the money economy. After all, they have done very well with cattle, sheep, and goats in the last 2,000 years or so. With better marketing facilities (the closest major cattle market is 50 miles away) and help with improving their stock, they could be important suppliers of beef to the nation.

Allowing and encouraging the Maasai to fulfill their aspirations in a way that incorporates their knowledge and many of their traditions is essential. Without the tolerance of the Maasai, the elephants have little chance of surviving, much less of living a relatively undisturbed existence in a naturally functioning ecosystem. The Maasai are a part of that ecosystem, and even though they are changing I believe some sort of mutually beneficial coexistence can be achieved.

The third major threat to Amboseli's elephants, possibly even an immediate one, is culling. "Culling" or "cropping" is the killing of a certain proportion of a population of animals for goals ranging from a simple reduction of numbers to full-scale game ranching at maximum sustained yield. The question whether to cull or not arises from and is answered by man's view of nature, natural resources, and conservation. On the whole man places himself above nature and thinks of nature as there to be exploited for the benefit of mankind as man thinks best. Exploitation may be consumptive (killing and eating animals, felling trees for timber, harnessing water for energy) or nonconsumptive (aesthetic appreciation of nature for recreation, scientific study). Man decides and controls.

The conservation of nature is generally accepted as a positive goal. Most plant and animal species (with the exception of those species that are inimical to man—weeds, tsetse flies, locusts, mosquitoes) are considered worth saving from extinction, either for ethical reasons or for the more utilitarian view that we do not know what value they might have for man in the future. To save species and their habitats, land has been set aside for conservation in countries throughout the world and conventions have been signed to prevent the overexploitation of fish and mammals in the sea.

Africa is acknowledged as one of the few remaining places on earth where large areas of wilderness containing a great diversity of plant and animal life still exist. No other continent has its variety and abundance of large mammals. For these reasons Africa has been a focus of conservation activity. In most African countries, parks and reserves have been created, many of them after the nation gained independence. Over the years growing numbers of people from more developed countries in both the East and the West have visited Africa's protected areas to see the wildlife and to benefit in some way from getting "close to nature."

For the people involved in the creation and maintenance of protected areas, there is always the problem of how to manage the park or reserve. What and how is man supposed to protect? The demarcating and gazetting of a protected area with boundaries and in some cases fences or ditches brings about some change no matter how big the area is. Is the area intended to be kept as it was when set aside, a capsule in time and space of a "representative fauna and flora"? Should natural animal and vegetation cycles be allowed to be played out? Should tourism be the prime objective? Or conservation? Or scientific research? Or maximum total revenue from all possible sources? The policies toward protected areas in Africa run the gamut of attitudes from preservation to consumptive utilization, and the methods employed range from hands-off nonintervention (the so-called laissez-faire approach) to intensive management.

At the laissez-faire end of the spectrum are those who believe that Nature (always with a capital) knows best, and left to herself she will take care of the habitats and the animals. Toward the other end of the spectrum is the attitude that once man has stepped in and created artificial boundaries, stopped hunting by local people, and built roads and hotels, a place is no longer natural and it can no longer take care of itself. Ecological processes have been disrupted, and animal and plant populations are no longer capable of self-regulation within acceptable limits. It is then man's responsibility to control and manage that protected area by regulation of animal numbers, burning schemes, the introduction of artificial watering points, and so forth. Somewhere in the middle but definitely closer to the laissez-faire view are those ecologists who see ecosystems as complex collections of interacting organisms shaped under the forces of natural selection, which they are just beginning to understand. In their view ecosystems show considerable resilience in the face of disturbance; and in the absence of biological understanding, management "tampering" in natural or semi-natural systems very often fails to have the desired effect or, in some cases, may make matters worse.

With such a broad range of views on how to manage parks and reserves, it is not surprising to find that the topic has generated a great deal of controversy, of which Kenya has had more than its share. The Kenya government policy regarding national parks does not dictate intensive management, but on the other hand the policy is open to a variety of interpretations. Sessional Paper No. 3 of 1975, "Policy on Wildlife Management in Kenya," states that national parks are managed exclusively for four objectives:

1. To preserve in a reasonably "natural" state examples of the main types of habitat which are found within Kenya for aesthetic, scientific and cultural purposes;

2. To provide educational and recreational opportunities for Kenyans;

3. To provide an attraction for tourists and so serve as a major basis for Kenya's economically profitable tourist industry; and

4. To sustain other activities not in conflict with the above.

The paper also states that agriculture, pastoralism, forestry, and consumptive wildlife uses are not allowed in national parks.

Nevertheless, the terminology is imprecise enough to allow arguments for the killing of animals in order to preserve the habitats in a "natural" state. And it is in this context that the long-standing debate about elephants and trees comes up. Elephants feed on trees and in doing so they sometimes kill trees or alter their form of growth markedly. This is generally viewed by people as being "bad." In some areas elephants have reduced woodlands to open grassland and bushland.

The classic examples are Murchison Falls National Park in Uganda and Tsavo National Park in Kenya. In Murchison Falls thousands of elephants were killed in the 1960s in a culling scheme to halt habitat change. (In the next decade over 90 percent of the remaining elephants were slaughtered by the army and poachers, but no one could have predicted that at the time.) In Tsavo a team of scientists, led by Richard Laws, was recruited in the 1960s to assess the elephant-tree situation and in order to do so they received permission to kill a sample number of elephants to determine their population dynamics. As a start 300 were killed, and the results suggested to the team that another 3,000 should be killed in order to gather enough data to determine how many more should be killed to reverse the woodland changes. The Kenya parks authorities felt that the killing of so many elephants for scientific purposes was abhorrent and in conflict with the stated use of the park and refused permission. The repercussions and the controversy surrounding that decision are still being felt today.

There were dire predictions that Tsavo would become a desert. Elephant numbers had built up both through immigration from surrounding areas where elephants were being pushed out by human expansion and through the elephants' own reproduction within the park. When Laws and his team were brought in there were estimated to be 40,000 elephants in the 7,000-square-mile ecosystem. After they left, some natural and some man-made events in the next decade changed the picture drastically. Droughts killed a few thousand elephants, and intensive poaching killed many more. By the late seventies only 8,000 to 10,000 elephants remained. Since then, Tsavo's vegetation has changed again, with bushes and trees returning throughout the park.

Neither side in the laissez-faire vs. culling controversy was happy with what happened in Tsavo. The noninterventionists, while encouraged by the evidence of habitat regeneration and resilience, did not see lawlessness and poaching as a solution. The intensive-management side viewed the outcome as a terrible waste. If the government, instead of drought and poachers, had culled the elephants, the country, including local people, would have reaped the benefit of one of its natural resources. As it was, the meat rotted in the park and the ivory was smuggled out of the country.

But what if culling had been allowed in Tsavo? To reduce the elephants to the point where woodlands were maintained it very likely would have interferred with a natural vegetation cycle, which, for example, may proceed over centuries from grassland to bush to woodland and back to grassland, under the influence of rainfall, elephant numbers, fire, and other factors. In addition, culling would no doubt have prevented the elephants themselves from responding reproductively to the compression and high densities. Even with the knowledge available on elephant population dynamics in the sixties, it could have been predicted that the elephants would have started breeding faster as soon as the levels of competition were reduced by culling. (In culling schemes whole family units are killed in an effort to cause as little disruption to the remaining population as possible. Under this system families are left intact with their leaders and their knowledge. Therefore, their ability to breed would not be adversely affected, as it would be in a poached population, where the older matriarchs and adult females are killed in families throughout the population.) The Tsavo elephants would eventually have been back to their former numbers and there would have to have been another cull. The wildlife authorities would have found themselves in a vicious circle, or as my colleague Keith has put it: Like the Red Queen of *Through the Looking Glass,* the wildlife manager would be running as fast as possible just to stay in one place.

Unfortunately, the specter of culling looms over the heads of Amboseli's elephants. Over 90 percent of the mature *Acacia xanthophloea* trees in Amboseli have died because of salinity and a rising water table, possibly accelerated through damage caused by elephants feeding on the bark. However, it appears that the older trees would have died whether or not elephants were present. The current "problem" is that the elephants are feeding on the young acacias that are growing up in place of the old trees. Seedlings are snapped off shortly after they emerge and saplings are pushed over, fed on, and killed. As a result very few stands of young acacias exist in the park, causing some people, including Jonah Western, whose interest has always been in the whole ecosystem, to worry about the park's habitat changes.

The number of elephants in Amboseli is known to have remained relatively stable over the last 20 years. Approximately 650 elephants comprise the Amboseli population today—the same number as estimated in the late sixties and early seventies.

Probably that many elephants, possibly more, have been in the area for the last several hundred years. In those centuries, the vegetation in Amboseli has apparently gone through various long-term cycles. Amboseli was known to have been a treeless waste when the explorer Joseph Thomson passed through toward the end of the nineteenth century. The trees that have been dying in the past 20 years were probably established as seedlings in the early 1900s. Thus we know for certain of one cycle of tree die-off and regrowth in the last 100 years and we can speculate that these cycles may have been going on for hundreds of years.

Nevertheless, Jonah feels that natural cycles can no longer occur in Amboseli under the present conditions. He is disturbed that the elephants appear to be spending more and more time in the central park area and that they are less frequently migrating out to the bushlands surrounding the basin. By concentrating in the basin they are feeding more on the young trees than they otherwise would, thus preventing the woodlands from regenerating. Unless the elephants change their migration patterns or their numbers are reduced, Amboseli will suffer a sustained loss of its habitat and wildlife diversity. Jonah has not yet recommended culling. He feels strongly that efforts should be made to try to reestablish the elephants' former migration patterns in order to reduce elephant density in the park. He has suggested various methods to accomplish this goal, such as creating a disturbance in the park (for example, shooting over the elephants' heads), fencing off part of the swamps, and allowing livestock back into the park, all in an effort to force the elephants out of the central area.

Other voices have not been as patient and many people have approached my colleagues and me about the question of culling the Amboseli elephants. No doubt the elephants are having a profound effect on the young acacias, but for how long this particular effect will continue is very much open to question. To cull the elephants in order to allow the regeneration of acacia trees would be a drastic, unreasoned step, given our present lack of knowledge of ecosystem dynamics.

First of all, in order to have any effect, the reduction would have to be severe; at least half the elephants, probably three quarters, would have to be killed. Second, a high proportion of the family units that use the central area (where the former woodlands occurred) would have to be killed. I and others who have worked on the project predict that the peripheral elephants would move into the central area as soon as the others were gone, just as they do every wet season when the residents move out. Third, the Kilimanjaro elephants are waiting in the wings. They have become more habituated each year and I would be greatly surprised if they did not fill the vacuum once the dominant Amboseli families were eradicated. Amboseli is much safer than the unprotected forests on Kilimanjaro.

In the meantime the Amboseli females who were left would continue to breed and a combination of births and immigration would bring the population back to where it started in a relatively short time. Maybe this would give the trees time to get a foothold, but can it possibly be worth the disruption caused by the shooting of several hundred elephants in a tiny park the size of Amboseli? The park is famous for its relaxed and highly viewable elephants. Tourists would not be pleased to see piles of rotting carcasses and skeletons and very frightened or possibly aggressive living elephants. And the social structure of the population would be shattered by the loss of so many families. Each clan and most bond groups would be profoundly affected. And what if the trees did not recover? Would another 100 or 200 elephants have to be killed? Once intensive management is started there is no looking back. One is no longer dealing with anything resembling a naturally functioning ecosystem with built-in resilience to environmental fluctuation and negative feedback systems. Something very precious is lost along with the elephants that are killed.

Lake Manyara is the only place I know of in Africa where an "elephant problem" has been perceived and the situation simply left to take its own course in the absence of culling or intensive poaching. When Iain concluded his study in 1971, he predicted that all the mature *Acacia tortilis* trees would be killed by elephants in the next ten years if the destruction continued at the then-present rate. At the time there was some regeneration, but not enough to replace the mature trees. Iain did not rule out culling but rather recommended the extension of the park. Eventually a small area to the south was added to the park but the density of elephants in the acacia woodlands remained about the same. After Iain left, the elephant population rose for five years; but then it fell by about 100 elephants in 1976 because of disease, leaving approximately the same number of elephants as when he began his study. About this time Rick Weyerhaeuser carried out a study of elephant-vegetation inter-actions in Manyara and made some unexpected discoveries. First, the elephants had greatly reduced their rate of feeding on the mature acacias and the die-off had slowed down consider-ably; second, the regeneration of young acacias was nearly rampant. Today Manyara is thick with acacias ten to fifteen feet high. The "problem" was solved without human intervention, and a valuable lesson in woodland ecology was learned.

There are many small parks in Africa surrounded by agri-cultural activity where animals have to be fenced in and where a laissez-faire policy would be considered impossible. Neverthe-less, I very strongly believe that the goal of conservation should be to strive to maintain some areas where ecological processes can continue with the minimum possible influence by man. We still have so much to learn from these processes, knowledge that could affect our very existence in the future, that I think it is a moral necessity to conserve natural ecosystems wherever pos-sible.

We also have a great deal more to learn about elephants. To cull in Amboseli would destroy forever the long-term study of a relatively undisturbed elephant population. There are not many places left in Africa where this kind of study can be carried out, and no other place with free-ranging elephants whose lives have already been followed for 14 years. This detailed record of the demographics of an entire population through time is unique, and it becomes more valuable as each year passes. The opportunity of developing a true understand-ing of elephant population dynamics from the level of the individual upward is still possible in Amboseli. Those of us who have been involved in the project are concerned to see it continue for at least another 20 years, and even then it would not cover the full lifespan of an elephant.

Against this background of the various threats to and prob-lems surrounding elephants, the Amboseli Elephant Research Project *is* going on. The camp is still standing, a lovely, peaceful oasis despite the loss of the trees. As a result of the kind generosity of friends, acquaintances, and even strangers, I raised enough money in 1985 to buy a new tent and a new Land-Rover. But the best thing that has happened to the project is the addition of two Kenyan assistants, both young women in their early twenties: Norah Njiraini and Soila Saiyie-lel. Norah has been with Joyce for two years and Soila, who began working for Sandy in July 1986, is now employed by me. They are remarkable in a country that has some fairly traditional views about the role of women. They have been trained by us to carry out the full range of field work. Joyce sent Norah to driving school in 1986, and until Soila joined us, Norah often went out by herself in a four-wheel-drive vehicle over rough terrain, collecting data on musth males for Joyce. She got charged by huge bulls, got stuck in the mud, had punctures and breakdowns, but she handled it all with great calm and a growing self-confidence that was rewarding to watch. Soila has learned to identify all the adult females and most of the calves faster than anyone else I know and while Sandy was away she did the censusing of the families and all the radio-tracking of two females who had been fitted with radio collars. Just recently I sent Soila to Nairobi to get her driver's license as well. Together Norah and Soila are an efficient team. And they do the dreaded vegetation plots and never complain. What makes me happiest is that they love their work and they love the elephants and are as taken up with their lives as any of us have been.

As for the main studies, Sandy Andelman has finished her two-year study of female-female cooperation and competition. Her preliminary results show that there is a dominance hierarchy among the family units in the Amboseli population and that factors such as the size of the family, the size of the bond group and the age of the matriarch influence dominance ranks. The distribution of the families in the dry season appears to be determined by both tradition and the relative dominance rank of the families. Sandy has recently started a comparative study of the Lake Manyara elephants, something both Iain and I have wanted to see for a long time.

Joyce is continuing her study of musth. Her original work generated excitement among all those interested in the relationship between hormones and aggression. She now has a major grant from the U.S. National Institute of Mental Health to explore this topic further, as well as funds from the Harry Frank Guggenheim Foundation, which is concerned with human dominance, violence, and aggression. Both organizations feel that her study can shed light on human behavior.

In addition, Joyce has undertaken a study of elephant communication. I find this project particularly exciting. As I mentioned before, Joyce waited for Phyllis and Keith to finish their studies before she began her research in late 1984. By luck, at the same time, we were approached by Katy Payne about vocal communication among elephants. Katy had worked for 12 years on whale communication, concentrating on the songs of the humpback whale, and wanted to switch to elephants. She chose to study Asian elephants, knowing that that field was less crowded with other researchers. She asked if she could come and observe us at work for a month in order to learn about elephants and how to work around them. She also offered to bring her powerful, highly sensitive recording equipment with her. We agreed and she came out for a month in January 1985 and then for another month in February 1986.

In between contacting us and arriving in Kenya for the first time, Katy made a very exciting discovery. She had gone to the Washington Park Zoo in Portland, Oregon, to observe a group of Asian elephants. At first when she was in the enclosure with the keeper she was disappointed because the elephants appeared to vocalize very infrequently. However, she soon detected a throbbing sensation in the air, although she could still hear no sound. Katy's experience with whales suggested to her that the elephants might be making sounds below the range of human hearing. She went back to Cornell University, where she is based at the Laboratory of Ornithology, arranged to obtain special recording equipment, and returned to Oregon with two fellow workers. After recording both in the daytime and through two entire nights, they found that the elephants were communicating with very low frequency calls, most of which could not be heard by the human ear. The special tape recorder was picking up the sounds but Katy and the others could not hear them unless the tape was speeded up ten times.

Human beings can hear sounds in the frequency range from 20–20,000 hertz (a unit of frequency equal to one cycle per second), and only people with very sensitive hearing can detect sounds down to 20 hertz. Some of the sounds the elephants were making ranged in frequency from 14 to 24 hertz and the sound pressure levels were high, at 85 to 90 decibels. (Human conversation is at about 65 decibels; amplified rock music is around 120 decibels.) The elephants were making very low but relatively loud sounds. Sound below the range of human hearing is called infrasound, that above ultrasound. While considerable work has been done on high-frequency sounds made by such mammals as bats, shrews, and porpoises, the elephant was the first terrestrial mammal reported to use infrasound.

When Katy came to Amboseli she and Joyce spent a month recording the elephants. The analysis carried out back at Cornell revealed that free-ranging African elephants also produced vocalizations with infrasonic components. Joyce and I had both suspected that something of this nature was occurring. We had seen instantaneous coordinated behavior of a group of elephants which made us think that one animal had given an alarm call that we could not hear. Katy and Joyce's recordings suggested that elephants could easily make an infrasonic alarm call, but what was more exciting was that they showed that elephants have a large repertoire of low-frequency calls used in a variety of contexts.

The discovery of the use of infrasound by elephants was fascinating in itself, but it also opened up a vast array of new interpretations of their behavior. Joyce has gone on to work with her own equipment and is primarily interested in the content of the vocalizations—that is, the messages that are being sent. It is known that low-frequency sounds travel over far greater distances and are less affected by trees and bushes than higher sounds. Theoretically some of the sounds that have been recorded in Amboseli (at up to 115 decibels) can carry for six miles, which could explain the coordination of movement and behavior of separated elephant groups. Contact calls and answers no doubt help elephants within families and bond groups find each other. These and other vocalizations may also enable elephants to avoid each other. In the intensive radio-tracking study that Rowan Martin carried out in Zimbabwe, he found that families from different clans would often be moving toward one another and then when they were two or three miles apart, they would change course in what appeared to be deliberate avoidance. In many cases the wind was not in the right direction to reveal each other's presence by scent. Low-frequency sounds may have alerted them.

Another important function of long-distance communication may be in finding mating partners. Almost inevitably after a female has been mated she gives a long, loud series of post-copulatory rumbles. These vocalizations contain infrasonic components. Joyce speculates that they carry over long distances and thereby attract males to the estrous female, who would then have a greater choice as to whom she mates with during the remainder of her estrous period. Males in musth make a characteristic rumble which is both very low and very loud. Joyce thinks that this rumble communicates different messages to different animals. It may attract estrous females and warn subordinate males to give a wide berth to the musth male. Often several males are in musth at the same time, and Joyce has noted that they manage to crisscross the park in search of females and yet usually avoid direct confrontations with each other. Infrasonic communication may very well be the key element in this spatial arrangement.

The observation of long-distance communication among elephants that I find most amazing and most disturbing was made in Zimbabwe. Garth Thompson, a wildlife guide, was for several years in charge of a private wildlife sanctuary adjacent to Hwange National Park. In the sanctuary were about 80 elephants whom he watched every day and got to know well. Generally they stayed in the area around the lodge he worked from and visited the artificial water holes that were constructed to enhance viewing. They were very relaxed, habituated elephants, and tourists in open vehicles could get as close to them as we do in Amboseli.

Next door in Hwange a very clear-cut policy had been set down to reduce the total number of elephants by culling in order to maintain tree densities. Intensive culling has been carried out each year for several years, with at least 1,000 and up to 4,000 elephants killed in one dry season. A family is herded by helicopter or light aircraft toward waiting marksmen and all the members, except calves of about one to three years old, are shot and killed in a matter of a couple of minutes. The calves are captured for sale to zoos and safari parks. It would be a clean, quick operation except for the babies, who scream and mill about and climb over their dead relatives in an effort to find and stay with their mothers. This system of culling is thought to least disturb an elephant population and is used in preference to shooting a few animals from each family. No doubt it is less disturbing, but the elephants know very well what is going on and that knowledge appears to be communicated throughout the population. In his last year in the sanctuary, Garth's 80 elephants disappeared on the very day that culling started in the park 90 miles away. Several days later he found them bunched together in the opposite corner of the reserve as far away from the park boundary as they could get. Somehow the message of danger and death had been relayed to them across those miles.

The descriptions of culling in Zimbabwe and Iain's reports on the rapid decline of elephants because of poaching made me realize how truly lucky Amboseli's elephants are. They must be some of the last elephants on earth who have the freedom and the relative safety to live their lives to the fullest, surrounded by their relatives from grandmothers down to the tiniest babies, dynamically and richly interacting with one another and their environment.

But I wonder how much longer it will be before the heavily armed, organized poachers turn to Amboseli. They would begin by shooting all the fine big bulls, then the younger bulls, and then the matriarchs and the older females one by one. In the last few years elephant poaching in Kenya has once more become rampant, particularly in the eastern part of the country. Once again Tsavo is under siege; and the very latest estimates for that park indicate that there may be no more than 5,000 elephants left from the original 40,000. In Amboseli there is not a trained field force capable of stopping poachers with automatic weapons. It is thought that the only reason they have not yet come is the presence of tourists and researchers.

And if the poachers do not come, what of the cullers? I feel sick when I think of a team of marksmen, skinners, and butchers moving into Amboseli and slaughtering whole families along with all their knowledge, their traditions, and their memories. The Ts, being a central family who use the acacia woodlands, would be among the first to be annihilated. It is almost too horrible to think about, but it is burying my head in the sand not to. Just the other day a university group to which Joyce was lecturing told her that they had heard that Amboseli's elephants were going to be culled. I am sure this is not true, but a kind of momentum begins to build as people see the aesthetically unappealing scene of dead and dying trees and think that "something" should be done. Killing the elephants seems the simplest and most direct solution, but only to people who have not watched individuals over 14 years; have not seen elephants greet one another with trumpets of joy; seen elephants, adults and calves alike, running and playing across an open pan in the moonlight; seen elephants trying to lift and hold up a stricken companion; seen a female stand by her dead baby for four days; or seen a seven-year-old calf gently fondle and stroke and feel the jaw of his dead mother.

Elephants are not so many rodents to be exterminated; they deserve something better than that and I am not afraid to say that ethics and morality should be essential considerations in our decisions for their future. Preserving habitats and trees and maintaining species diversity are important goals in conservation, but nothing will convince me that killing Slit Ear, Tallulah, and Tuskless and their families would be worth the achievement of those goals. The world would not be a better place if they were dead.

In the last two years I have realized that more than anything else, more than scientific discoveries or acceptance, what I care about and what I will fight for is the conservation, for as long as possible, not of just a certain number of elephants, but of the whole way of life of elephants. My priority, my love, my life are the Amboseli elephants, but I also want to ensure that there are elephants in other places that are able to exist in all the complexity and joy that elephants are capable of. It may be a lot to ask as we are about to enter the twenty-first century, but I think it is a goal worth striving for.

Epilogue

December 26, 1986

It was the morning after Christmas. Several friends had come to Amboseli to celebrate the holiday with me. We had had a huge feast and to everyone's enjoyment over 100 elephants had arrived and fed in the glades around the camp. The ones who came right into the camp were, of course, Tuskless, Tonie, Tilly, their calves, and Tom, and later Toby, Tim, and Right Fang. Peter said they had come to say hello to me, that no elephants had been in the camp for days. Part of me had wanted to believe him, and I had felt a small warm glow of happiness.

On this morning I drove out from the camp with one of my friends at about 8:00 A.M. He is a very keen and knowledgeable ornithologist and we had decided to go bird-watching. (He promised to teach me to identify the waders who migrate to Amboseli from Europe every winter and who drive me to distraction when I cannot tell a green sandpiper from a wood sandpiper or a greenshank from a marsh sandpiper.) We were on our way to the middle of Longinye swamp where there were open pools of water, but, of course, because I was *not* looking for elephants we kept coming upon them.

There had been good rains in November and December and the park was green and lush. The elephants we met walked with spirit and energy after nearly two months on a good diet. The contrast to their lethargic demeanor at the end of the long, hard dry season struck me yet again.

The first elephants we saw were peripheral families—the BCs, HBs, and JBs—who had apparently made a beeline to the central swamps once the rains had come, knowing that the central elephants either would be gone or would tolerate them. I did not stop to census them but drove on. We next saw the BBs, whose matriarch, the improbable and wonderful-looking Big Tuskless, had sadly died earlier in the year. I noted that everyone was there plus their close bond-group members, the UAs, led by Ulla. I could not yet decide whether Barbara or Bette (another tuskless) was taking over the role of matriarch. Then we came upon a group just emerging from a patch of *suaeda* bushes. I saw that it was the TCs and TDs and was tempted to keep going but stopped because Tallulah was hurrying, running, and I thought they might have been disturbed by something. They were not. Tallulah, usually acting the dignified matriarch these days, was playing. My notes read:

> 0850 Hurrying along on S. side, covered in mud, Tallulah and her two calves and Theo's '82. These four are being silly—floppy running, then beating through *suaeda* bushes, pirouetting around, heads up, tails curled. Theo stays calm. Tallulah picks up a log and throws it, then backs over to it and kicks it. Tao just as silly.

> 0855 Behind them come Slit Ear and family, more dignified. At the back is Tess with a brand new baby—still pink behind the ears but walking well. Would guess it was born yesterday, on Christmas day.

My visitor was delighted. He had not realized that adult elephants played like that. I was equally delighted, for more complicated reasons. It was the second time I had seen the TD females playing that year. On an earlier occasion I had watched

Theodora cavorting in the swamp in about five feet of water. She smacked the water and swished her trunk about, splashing and making waves. She sat down and raised her head into the air and flopped over on her side, submerging herself completely before finally coming up with vegetation strewn about her head and tusks. Then she started gently sparring with little Slo and they chased each other through the muddy water and thick vegetation for another ten minutes. That time and again on this day I felt a surge of joy for them. Tallulah and Theodora, after more than two years, seemed to have recovered from Teresia's death, and they were once again the sprightly, often whimsical elephants that I remembered and loved so much. They were always my favorite young females, as much as I tried to be impartial. Seeing them healthy and active and playful gave me hope for their future and for the future of the other Amboseli elephants as well.

And Tess's little calf with her bright-pink ears and tiny wiggly trunk was the future right there. Would she grow up in relative peace and freedom? Would she have the tremendous good fortune to have her mother and even her grandmother around in 13 or 14 years' time, when she had her first calf? I hoped so, and on a beautiful clear bright morning with the snows of Kilimanjaro glistening in the sunlight, I believed it was possible.

A few days later, at the end of 1986, with a mixture of both hopes and fears for them, I recorded the status of the Ts. (In keeping with our policy of naming calves when they reached four years old, I gave the four '82 calves names. I called Tonie's calf Truman after a botanist friend of ours, Tara's calf Turner at the request of one of the donors to the project, Tess's calf Tamsin after another friend, and Theo's calf Tebald for no particular reason except the first three letters made a unique code. At nearly five years old Tamsin still had no tusks and so the Ts had another tuskless.) In all, the three families in the bond group consisted of 22 animals, close to the same number as when I first saw them 13 years before, in 1973. There were also five T males out in the population biding their time until they could start siring offspring: Teddy, Tim, Tolstoy, Right Fang, and Toby. Most of the original adults were gone, but their descendants were definitely carrying on:

TA	Tuskless (62)	♀
	C'84	♀
	Tuo (80)	♂
	Tonie (67)	♀
	Truman (82)	♂
	Tilly (70)	♀
	C'83	♂
	Tom (79) [ex-TB]	♂
TC	Slit Ear (36)	♀
	C'83	♀
	Slo (80)	♀
	Tabitha (76)	♀
	Tara (67)	♀
	Turner (82)	♂
	Tess (62)	♀
	C'86	♀
	Tamsin (82)	♀
TD	Tallulah (63)	♀
	C'83	♀
	Tao (80)	♀
	Theodora (67)	♀
	Tebald (82)	♂

Amboseli Elephant Research Project Publications

Croze, H. J., Hillman, A.K.K., and Lang, E. M.
1981. Elephants and theireach other. *Dynamics of Large Mammal Populations* (ed. C. W. Fowler). New York: John Wiley, 297–316.

Douglas-Hamilton, I., Hillman, A.K.K., and Moss, C. J.
1981. Notes on vertical photography of elephants for age determination. *ILCA Monographs*, 4:131–142.

Lee, P. C.
1986. Early social development among African elephant calves. *National Geographic Research*, 2:388–401.

1987. Allomothering among African elephants. *Animal Behaviour*, 35:278–291.

In review. Family structure and its effects on female reproductive effort. *Comparative Socioecology of Mammals and Man* (eds. V. Standen and R. Foley). Oxford: Blackwell Scientific Publications.

Lee, P. C., and Moss, C. J.
    1985. Early maternal investment in male and female African elephant calves. *Behavioural Ecology and Sociobiology,* 18: 353–361.

Lindsay, W. K.
    1982. Habitat selection and social group dynamics of African elephants in Amboseli, Kenya. M.Sc. thesis, University of British Columbia, 200 pp.

    1983. Elephants, trees, and people. *Wildlife News,* 18(2): 8–11.

    1986. Elephant problems and human attitudes. *Swara,* 9(3): 24–27.

    1986. Trading elephants for ivory. *New Scientist,* 112(1533): 48–52.

    In review. Food intake rates and habitat selection in elephants in Amboseli, Kenya. Proceedings of the International Symposium on African Wildlife. Uganda Institute of Ecology, Kampala.

    In prep. Feeding behaviour and ecology of African elephants in Amboseli National Park, Kenya. Ph.D. thesis, University of Cambridge.

Lindsay, W. K., and Mungai, M. E.
    In prep. A test of the dropping count method for estimating elephant density in Amboseli National Park, Kenya.

Moss, C. J.
    1977. The Amboseli elephants. *Wildlife News,* 12(2):9–12.

    1978. A family saga. *Swara,* 1(1):34–39.

    1980. What do you do with a 300-pound nose? *International Wildlife,* May-June, pp. 36–39.

    1981. The two and only. *Animal Kingdom,* 83(6):25–27.

    1981. Social circles. *Wildlife News,* 16(1):2–7.

1982. *Portraits in the Wild: Behavior Studies of East Africa Mammals* (second edition—revised). Chicago: The University of Chicago Press, 371 pp.

1983. Oestrous behaviour and female choice, in the African elephant. *Behaviour,* 86(3/4):167–196.

In prep. Flexibility in the social system of the African elephant.

In prep. The demography of the elephant population of Amboseli National Park, Kenya.

Moss, C. J., and Poole, J. H.
   1983. Relationships and social structure of African elephants. *Primate Social Relationships* (ed. R. A. Hinde), Oxford: Blackwell Scientific Publications, 315–325.

Payne, K. B., Poole, J. H., and Langbauer, Jr., W. R.
   In prep. Infrasonic calls in free-ranging African elephants: a possible long distance communication network.

Poole, J. H.
   1982. Musth and male-male competition in the African elephant. Ph.D. thesis, University of Cambridge, 158 pp.

   1987. Elephants in musth, lust. *Natural History,* 96 (II): 46–55.

   1987. Raging bulls. *Animal Kingdom,* 90 (6): 18–25.

   In press. Rutting behaviour in African elephants: the phenomenon of musth. *Behaviour.*

   In press. Honest advertisement of intent: the aggressive state of musth in African elephants. *Animal Behaviour.*

   In prep. Guarding behaviour in African elephants.

Poole, J. H., Kasman, L. H., Ramsay, E. C., and Lasley, B. L.
   1984. Musth and urinary testosterone concentration in the African elephant (*Loxodonta africana*). *Journal of Reproduction and Fertility,* 70: 255–260.

Polle, J. H., Payne, K. B., Langbauer, Jr., W. R., and Moss, C. J.
In review. A sample of very low frequency calls from free-ranging African elephants: their social context and possible role in long distant communication.

Poole, J. H., and Moss, C. J.
1981. Musth in the African elephant, *Loxodonta africana. Nature,* 292: 830–831.

Western, D., and Lindsay, W. K.
1984. Seasonal herd dynamics of a savanna elephant population. *African Journal of Ecology,* 22: 229–244.

Western, D., Moss, C. J., and Georgiadis, N.
1983. Age estimation and population age structure of elephants from footprint dimensions. *Journal of Wildlife Management,* 47(4): 1192–1197.

Young, T. P., and Lindsay, W. K.
In press. Role of even-age population structure in the disappearance of *Acacia xanthophloea* woodlands. *African Journal of Ecology.*

Index

About the Author

After graduating from Smith College and working as a reporter/researcher for *Newsweek*, Cynthia Moss moved to Africa in 1968. Since then, while holding a variety of jobs and titles, she has devoted herself to the study of elephant life. Currently Director of the Amboseli Elephant Research Project, Senior Associate of the African Wildlife Foundation, Kenya, and Research Associate at the Sub-Department of Animal Behavior at Cambridge University, she has published numerous scientific papers and popular articles in magazines such as *Smithsonian* and *Nature*; received grants from the African Wildlife Foundation and the New York Zoological Society; and is the author of *Portraits in the Wild*. In 1985 she received the Smith College medal for alumnae achievement. Her work has been featured on the half-hour *Wild Kingdom* TV show and on ABC news in a Roger Caras report; and she has been interviewed and filmed by David Hartman for ABC's *Good Morning America*.